D0342295

RECEIVED

NO LONGER PROPERTY OF
SEATTLE PUBLIC LIBRARY

Francis Crick

Eminent Lives—brief biographies, by distinguished authors, of canonical figures—joins a long tradition in this lively form, from Plutarch's *Lives* to Vasari's *Lives of the Painters,* Dr. Johnson's *Lives of the Poets* to Lytton Strachey's *Eminent Victorians.* Pairing great subjects with writers known for their strong sensibilities and sharp, lively points of view, the Eminent Lives are ideal introductions designed to appeal to the general reader, the student, and the scholar. "To preserve a becoming brevity which excludes everything that is redundant and nothing that is significant," wrote Strachey: "That, surely, is the first duty of the biographer."

BOOKS IN THE EMINENT LIVES SERIES

Robert Gottlieb *on* George Balanchine

Christopher Hitchens *on* Thomas Jefferson

Paul Johnson *on* George Washington

Michael Korda *on* Ulysses S. Grant

Edmund Morris *on* Beethoven

Francine Prose *on* Caravaggio

FORTHCOMING BOOKS

Karen Armstrong *on* Muhammad

Louis Begley *on* Franz Kafka

Bill Bryson *on* William Shakespeare

Joseph Epstein *on* Alexis de Tocqueville

Ross King *on* Machiavelli

Peter Kramer *on* Sigmund Freud

Brenda Maddox *on* George Eliot

GENERAL EDITOR: JAMES ATLAS

ALSO BY MATT RIDLEY

The Agile Gene: How Nature Turns on Nurture

Genome: The Autobiography of a Species in 23 Chapters

The Red Queen: Sex and the Evolution of Human Nature

The Origins of Virtue:
Human Instincts and the Evolution of Cooperation

FRANCIS CRICK

Discoverer of the Genetic Code

Matt Ridley

EMINENT LIVES

FRANCIS CRICK. Copyright © 2006 by Matt Ridley. All rights reserved. Printed in the United States of America. No part of this book may be used or reproduced in any manner whatsoever without written permission except in the case of brief quotations embodied in critical articles and reviews. For information, address HarperCollins Publishers, 10 East 53rd Street, New York, NY 10022.

HarperCollins books may be purchased for educational, business, or sales promotional use. For information, please write: Special Markets Department, HarperCollins Publishers, 10 East 53rd Street, New York, NY 10022.

Title page image copyright A. Barrington Brown/Photo Researchers, Inc.

FIRST EDITION

Designed by Elliott Beard

Printed on acid-free paper

Library of Congress Cataloging-in-Publication Data

Ridley, Matt.

Francis Crick : discoverer of the genetic code / Matt Ridley.—1st ed.

p. cm.—(Eminent lives)

Includes bibliographical references.

ISBN-10: 0-06-082333-X

ISBN-13: 978-0-06-082333-7

1. Crick, Francis, 1916– 2. Geneticists—Biography. 3. Molecular biologists—Biography. 4. Genetic code—Research—History. 5. DNA—Research—History. I. Title. II. Series.

QH429.2.C75R53 2006

576.5'092—dc22

[B]

2005055878

06 07 08 09 10 ID/RRD 10 9 8 7 6 5 4 3 2 1

For Felicity

Contents

Prologue:	Life Itself	1
Chapter One:	Crackers	3
Chapter Two:	Three Friends	17
Chapter Three:	Cambridge	29
Chapter Four:	Watson	45
Chapter Five:	Triumph	59
Chapter Six:	Codes	77
Chapter Seven:	Brenner	97
Chapter Eight:	Triplets and Chapels	115
Chapter Nine:	The Prize	127
Chapter Ten:	Never in a Modest Mood	145
Chapter Eleven:	Outer Space	163
Chapter Twelve:	California	179

Chapter Thirteen: Consciousness 191

Epilogue: The Astonishing Hypothesiser 207

Sources and Acknowledgements 211

Prologue

Life Itself

ON 8 JUNE 1966 THERE was a party on the lawn of Blackford Hall, one of the houses belonging to the famous Cold Spring Harbor Laboratory on the north shore of Long Island. After a lobster banquet, a girl called Fifi, wearing a bikini, burst out of a giant cake—not the usual climax to a scientific conference. But it was not the usual pretext for a party. It was the fiftieth birthday of Francis Crick, a man routinely described by other scientists "the cleverest person I have ever met." It was also the actual birthday of Crick's scientific baby, the genetic code. Crick had just put the finishing touches on a little cipher chart, which gave the exact protein translation for all but one of the three-letter words in the lexicon of DNA. That code, he rightly suspected, was universal to all living creatures—proving that all livings things have a single common ancestor. It was, in fact, the reason they were living. It carried messages from the past to the future, messages about how to build living bodies from food by directing the synthesis of proteins. It was the very definition of the difference between living and nonliving, a difference Crick had set out deliberately to define nearly 20 years before.

On that day Crick stood on top of the scientific world. Others had done some of the crucial experiments in the decoding of the code, and others had shared the excitement of vital discoveries along the way—the messenger, the adaptor, the triplet nature of the code—but Crick had been there at every step, the dominant theoretical thinker, the best guesser, the indefatigable sceptic, the loudest debater, the conductor of the scientific orchestra. Thirteen years before, he and James Watson had discovered suddenly and famously that there was a code at all, when they stumbled on the structure of DNA, the stuff from which genes are made. Now, the code was cracked. The reason a rabbit differed from a rock was that it had a long message inside it written in three-letter words in a four-letter alphabet. The reason a rabbit differed from a person was that it had a different sequence of letters in its message. Life was that simple. Crick could cross it off a mental list he had imagined long ago—a list of mysteries that must be snatched from the hands of mystics and handed over to reason. It had been a short list, with just two items: life and consciousness. Life was done.

As a small boy Francis Crick had been haunted by a fear that by the time he grew up everything would have been discovered. Inspired by Arthur Mee's *Children's Encyclopedia*, the boy had become fascinated by the unexpectedness of scientific answers. From a very early age he longed to find some of his own, but would there be anything left? "Don't worry, ducky," said his mother. "There will be plenty left for you to find out."

Chapter One

Crackers

FRANCIS HARRY COMPTON CRICK was born on 8 June 1916, at the height of World War I. The day before he was born, the news had broken that Lord Kitchener, Britain's celebrated minister of war, had been killed on board a cruiser bound for Russia. When Crick was a few weeks old, the first day of the battle of the Somme would claim 20,000 British lives. Far away from all this death, Crick was born at home in Holmfield Way in Northampton, a middle-class street in a middle-size town in the middle of the English Midlands. He was the son of a shoe manufacturer, Northampton being the shoemaking capital of Britain. Its streets were full of workshops and factories where leather-aproned workers still hammered and stitched soles, heels, and uppers. Shoemaking was an increasingly mechanised trade, thanks partly to the invention of one Thomas Crick of Leicester, who in 1853 took out a patent for an improved method of fixing uppers to soles with tacks or rivets instead of stitches. But, perhaps fortunately for posterity,

Thomas Crick was no ancestor of Francis, who consequently was spared the distractions of great wealth.

Crick's Y chromosome had not wandered far in two centuries, or perhaps for much longer. Crick is not an uncommon surname in the Midlands, the village of Crick in Northamptonshire being its probable origin. In 1861 Francis's great-grandfather Charles Crick was a fairly prosperous farmer, employing 20 men and boys on his 231 acres at Pindon End farm near the lace-making village of Hanslope just 10 miles south of Northampton. Charles's second son, Walter Drawbridge Crick, born in 1857, took a job as a clerk in the goods department of the London and Northwestern Railway, whose track bisected his father's farm. He soon switched to working as a travelling salesman for a shoemaker called Smeed and Warren. In 1880, when he was just 22 years old, he joined two others to start his own boot and shoe factory: Latimer, Crick, and Gunn, at Green Street, Northampton. (The churchyard at Hanslope has several Latimers buried in it, as well as some Cricks, so perhaps Latimer was a family friend.) The business thrived and expanded to Madras in India. At one time it also had five shops in London, and later it made military boots for those doomed young men at the Somme. By 1898 William Latimer and Thomas Gunn had retired, leaving Walter Crick the sole owner of the firm. He did well enough to build a substantial stone mansion, Nine Springs Villa, on Billing Road on the eastern side of Northampton. But five years later Walter Crick (at age 47) died of a heart attack, leaving the firm in the hands of his widow, Sarah—who survived him by 31 years—and two of his four sons, Walter and Harry, who carried on the business until it failed during the Depression.

The original Walter's enthusiasm for shoes, lucrative though it was, seems to have come second to his passion for science, and for collecting—fossils, books, stamps, coins, porcelain, and furniture. His friends found him energetic and argumentative. Said one, in terms that might later have been applied to the grandson: "He was just as fond of springing a new and carefully stored fact into a discussion as he was of trumping a suit the first time round." He was an amateur naturalist of some local repute, who eventually wrote a two-part survey of the Liassic foraminifera of Northamptonshire and had two gastropods named after him. On foot and bicycle, he wandered the lanes of Northamptonshire collecting fossils and turning over rocks to look for snails. It was a tiny mollusc that caused Walter, grandfather of the greatest biologist of the twentieth century, to forge a brief link with the greatest biologist of the nineteenth: Charles Darwin.

It happened thus. On Saturday, 18 February 1882, Walter Crick was out hunting for water beetles (a curious occupation in winter, surely). We know this because later that day he wrote hesitantly to Darwin to report what he had found. "I secured a female Dytiscus marginalis," he told the great evolutionist, "with a small bivalve [cockle] that I think is Sphaerium corneum very firmly attached to its leg." Darwin replied three days later with a barrage of questions. He wanted to know the length and breadth of the shell, and how much of the leg (which leg?) had been caught; and he suggested a communication to the magazine *Nature*. To a young railway clerk turned shoemaker with (to judge by his handwriting) only a rudimentary education, this reply must have been a matter for some excitement. Crick replied with not only the answers, but also the beetle and the shell. Both arrived alive, so Darwin put the "wretched" insect in a bottle with chopped laurel leaves, "that it may die an easy and quicker death." He then sent both specimens off to an expert on shells for identification, but the expert was away and the specimens were returned, broken, by a servant. Meanwhile, Crick had returned to the same pond on a Sunday and found a dead frog with a cockle of the same kind attached to its foot. On 6 April, Darwin published a letter in *Nature* describing Crick's cockles, as a triumphant vindication of his long-held theory that peripatetic molluscs hitch lifts with other animals to get from pond to pond. It was to be Darwin's last publication: 13 days later, he died.

Walter and Sarah Crick had five children, born between 1886 and 1898. They were destined to grow to adulthood just as the relative peace and freedom of Edwardian England vanished, and they suffered their share of disappointments in the 30 years of war and slump that followed. The eldest, Walter, as senior director of the business, gets the family's blame for the failure of the shoe firm in the mid-1930s. One of the causes—or consequences—may have been his passionate interest in a new and somewhat conspiratorial theory of economics, championed by the chemist Frederick Soddy, a Nobel laureate, to the effect that economic crises were caused by banks' having less than 100 per cent cover for their liabilities. In 1939 Walter Crick even published a short book with Soddy urging the world to, in the words of the title, *Abolish Private Money or Drown in Debt*. Walter emigrated to America at the start of World War II and spent the rest of his working life as a sales agent for a rival shoemaking firm, eventually settling on an orange farm in California. The second son, Harry, was Francis's father. His role was to manage the shoe shops in London while Walter ran the factory. The factory's failure left Harry in such straits that he could not afford the boarding fees for his younger son at school. The third son, Arthur, did best. He avoided the family business and set up as a pharmacist in Kent, making antacid pills for indigestion and eventually owning a string of successful shops. This made him sufficiently wealthy to pay for his nephew Francis to stay at University College as a graduate student, a subvention that undoubtedly saved Francis from having to give up science. The fourth son, William, was killed at age 20 in 1917 at the battle of Arras, serving as a second lieutenant in the King's Own Yorkshire Light Infantry. The youngest child, Winifred, married Arnold Dickens, who owned a leather manufacturing company, and remained in Northampton, where she bore four children, retaining into old age a reputation for plain speaking.

Francis Crick never knew his naturalist grandfather, but his uncle Walter had an interest in science. Walter taught young Francis glass blowing and encouraged him in precocious and dangerous chemistry experiments in a garden shed. (Too successful explosions in sheds or attics feature in the early lives of many twentieth-century scientists; but unlike, say, the evolutionary theorist Bill Hamilton, Crick retained all his fingers.) Francis's father, Harry, a jolly man, was more interested in tennis, bridge, and gardening than science. He even played once at Wimbledon, though he lost his only match. Both his sons inherited his talent for and interest in tennis, Francis making the school tennis team and his younger brother Tony going on to play at the county level. But Francis had given the game up long before he met the tennis-mad James Watson.

Harry Crick married Annie Wilkins in September 1914. Like him, she was the child of a self-made businessman. She was one of five children of F. W. Wilkins, who had started a successful chain of clothing stores. But in his will Wilkins provided for the managers to buy all but one of the stores, so Annie and her sister Ethel inherited just that one, Wilkins and Darking, in the Wiltshire town of Trowbridge, the other siblings having sold their shares. The Wilkins money would later help to make Francis moderately well off, his aunt Ethel leaving him her substantial house in Cambridge when she died. Ethel was a teacher and Annie a nurse. They were both strong characters and, for a while, defiantly single. Obsessively interested in people's health, Annie was nearly teetotal; when she was told by a doctor to drink stout for her own health, she did so in bed, holding her nose. She was 35 when she suddenly married Harry Crick, 10 years her junior. She bore two sons, Francis and Tony. Family tradition has it that Aunt Ethel carried the newborn Francis to the top of the house to ensure that he would "rise to the top" of his profession.

Francis therefore grew up with the habits and memories of the well-off middle class, but without much of the money. As wealth evaporated for the Cricks in the difficult years of the twentieth century, so religion faded, too. The Cricks were, like the Darwins, of nonconformist Unitarian stock (the Unitarian church, with its reputation for doubt, was a good incubator of scientists), but they had deserted the Unitarians over a dispute with the local pastor and joined the Congregational Church on Abington Avenue in Northampton. Annie was also a Congregationalist. Harry Crick became secretary of the church, but neither parent was especially devout. Harry sometimes played tennis on Sunday afternoons, a fact that young Francis was told not to mention in front of other members of the church. The roots of Francis's fervent lifelong atheism cannot therefore lie in rebellion against having been coerced into belief as a child. Although his sudden refusal to go to church, at about age 12, briefly upset his mother, it does not seem to have resulted in much disturbance to anybody's equanimity. At boarding school, where church was not optional, singing in the choir and sitting through sermons was no penance, just part of the ritual. Francis found illogical sermons entertaining rather than upsetting.

He had no doubt that what caused him to lose his faith was science which revealed some of the assertions of the Bible as false. "And if some of the Bible is manifestly wrong," he wrote in his memoirs, "why should any of the rest of it be accepted automatically?" But this worked both ways: his loss of faith also motivated his science. "What would be more important than to find our true place in the universe by removing one of these unfortunate vestiges of early belief?"

The precocious atheism, the boyhood fascination with facts and science, the confident scepticism, a certain facility with mathematics—the signs of brilliance were all there. But, unlike his later collaborators Jim Watson and Sydney Brenner, who went to the university before they were 16, Crick was no prodigy. Indeed, for the first 35 years of his life he was, in achievement at least, unremarkable. Although in 1930 he won a scholarship from Northampton Grammar School to attend boarding school at Mill Hill in north London (the same nonconformist school his father and three uncles had attended), he did not especially shine there. He was "extrovert and mildly eccentric" and wore suede shoes, according to one school friend, who called him "Crackers." His headmaster was the first of many people throughout his life who were struck by his loud, peculiar laugh. He was on the school tennis team, and was good at mathematics and science, winning a chemistry prize; but he was never made a prefect, and he failed to get the expected place at Oxford or Cambridge. He did once give a talk on how the Bohr theory of the atom, plus quantum mechanics, explained the periodic table—but as he said later to Oliver Sacks, he was not sure how much of all that he really understood.

In 1934 he secured a place at University College, London (UCL), to study physics, and he emerged three years later with a disappointingly average second-class degree. His parents having moved to north London so that they could afford Tony's fees as a day boy at Mill Hill, he was able to commute to the university from the family home. UCL, nicknamed the "godless institution of Gower Street," had been founded in 1826 explicitly to provide education with no religious content. His best friend from undergraduate days was Raoul Colinvaux, with whom he later shared a flat and who eventually became a barrister. Financially supported by his uncle Arthur, who was prospering, Crick started a PhD at UCL under Professor Edward Neville da Costa Andrade, a former colleague of Ernest Rutherford who had written a much-read book *The Structure of the Atom*, and later became science correspondent of the *Times*. Short, impeccably dressed, and a witty though prickly conversationalist with a flair for poetry, Andrade sounds like the perfect professor for a young student with Crick's curiosity. But they do not seem to have gotten much out of each other. Andrade, whose research interest was in the mathematics of flow, viscosity, and creep, gave Crick the "dullest problem imaginable," to measure the viscosity of water under pressure between 100 and 150 degrees Celsius. Apart from some fun constructing the oscillating copper apparatus, and the fact that he actually won a prize for his work in his second year, the experience was, Crick recollected, "a complete waste of time."

Hermann Göring put the viscosity experiment out of its misery. At the outbreak of war in 1939, UCL's physics department was evacuated to Wales, but Crick chose to stay at home. He spent the first weeks of the war playing squash with his brother on the courts at the deserted Mill Hill school, the pupils having also been sent to Wales. By early 1940 he had found a civilian job in research in the Admiralty. Later in the war a land mine (a naval magnetic mine dropped by parachute from an aeroplane) scored a direct hit on the carefully constructed viscosity apparatus, somewhat to Crick's relief.

In the Admiralty, Crick was working under Harrie Massey, who was another professor at UCL and another student of Rutherford's. Massey, the son of an Australian gold prospector, was an expert on quantum mechanics and atomic collisions and had joined UCL in 1938 as Goldsmid professor of mathematics. Massey's job was to lead a small team first designing sweeps for magnetic mines and then designing magnetic mines that would defeat enemy countermeasures. The British had invented the first magnetic mine in 1917 but had failed to develop it between the wars, because of a dispute between the navy and the air force about whose toy it was, so they were shocked when the

Germans began seeding the Channel coast and the Thames estuary with these devices in the autumn of 1939. A magnetic mine sits on the seabed, armed (by the pressure of its descent) to go off when it detects a local disturbance to the earth's magnetic field. Steel-hulled ships concentrate the earth's magnetic field slightly in a direction downward from the north pole. So long as the water is fairly shallow, this concentration is enough to trip the mechanism of the mine when the ship is overhead. In November 1939, 200,000 tons of shipping were sunk by mines and the port of London was all but closed. Secretly, the British government was desperate. As far as the population was concerned, this was the surprisingly pleasant "phony war." But for a month the country was in real peril. Fortunately, on the night of 23 November, a German seaplane was seen dropping parachutes near Shoeburyness at high tide, and a quick-thinking coastguardsman realised that the objects would be exposed by the low tide at four o'clock the next morning. Sure enough, as the water retreated, two magnetic mines emerged. One was carefully defused, so that its mechanism could be examined. A magnetic needle slightly weighted at the south pole would dip to the north if a ship passed over, closing a circuit and allowing a charge to pass through a detonator. The scientists then went to work to design sweeps for the magnetic mines and protection for steel ships. It was for this that Crick was recruited. Electric cables towed behind two wooden-hulled boats soon proved effective at blowing up mines; and "degaussing"—the induction of a temporary magnetic field downward from the south pole in the hull of a steel ship using giant coils in a dock—soon made ships less vulnerable.

On 18 February 1940, Crick, who was now employed, married Doreen Dodd, a fellow undergraduate from UCL with a degree in English literature. She was tall and fair with a broad face and a taste for the novels of Tobias Smollett, and was working as a clerk at the Ministry of Labour. It was a low-key wartime wedding, without a honeymoon, at a registry office in St. Pancras. In November, at the height of the blitz, the Cricks' son, Michael, was born, inevitably during an air raid. Crick commuted to work at the Admiralty Research Laboratory at Teddington, until the team was later moved to the Mine Design Department's headquarters in a Regency country house called West Leigh, near Havant, on the south coast. Here his job shifted from countermeasures against enemy mines to designing mines immune to such countermeasures to be used against the enemy. He rented a dilapidated house nearby for Doreen and the baby. It cannot have been an easy time, as nearby Portsmouth was bombed almost nightly and the south coast was a military zone in which movement was highly restricted. Crick was a junior member of the Mine Design Department, but there are hints that his strong personality—which was not always loyal to the team—was noticed. Rank-conscious senior naval officers were not accustomed to being told by a scruffy young man in civilian clothes that they were talking nonsense. Despite his youth, Crick was made leader of a team code-named MX, with about a dozen men working under him. Their job was not to alter the explosive mechanism of the mines but to tinker with the circuits that triggered a mine, the better to defeat enemy countermeasures.

As the war went on, Crick found himself drawn farther into strategy and intelligence. One day a German sailor in a bar in an occupied port, relaxing a little too much, let slip that his ship carried a sort of huge magnet in its bow. Someone passed the remark to British naval intelligence. The ship in question, it emerged, was called a Sperrbrecher, and it was much larger and more heavily armed than most minesweepers. It housed in its bow a 500-ton electromagnet to detonate magnetic mines in front of it (a system that worked only against the British horizontal-field magnetic mines, not against the German vertical ones). Massey asked how Crick would counter such a vessel. Crick immediately suggested that an especially insensitive mine triggered only by a very strong magnetic field would detonate directly beneath the Sperrbrecher, rather than in front of it. But such a mine would be quite harmless to normal ships, and many naval officers could not at first see the point of laying mines that other enemy ships could pass over with impunity. Crick persisted. To carry out his plan, he needed to know the precise strength of the Sperrbrecher's magnet, and nobody could think how to obtain this information. Tipsy sailors were unlikely to know it, let alone repeat it. By good fortune one day in July 1942, a Royal Air Force (RAF) reconnaissance plane over Lorient photographed a Sperrbrecher just after it had exploded a mine. In two successive photographs the wake of the ship was cutting through the circular wash of the mine explosion, allowing Crick and his colleagues to calculate—from the speed of the ship, the depth of the water, the size of the mine, and the diameter of the wash—exactly how strong the magnetic field of the Sperrbrecher was.

A special mine, with a resistance across the relay in the trigger mechanism to reduce its sensitivity, was tested successfully off Portsmouth. The RAF then laid several such mines in the area patrolled by the very Sperrbrecher that had been photographed. Within two weeks it was on the bottom of the sea. By the end of the war, more than 100 Sperrbrechers had been sunk, not only leaving German waters vulnerable to mines but wasting a large tonnage of the enemy's expensive ships. Crick later repeated a similar trick with acoustic mines, triggered by the noise of a ship's engines, to which the German and Allied navies had both turned during the course of the war. His "special" mines, despite being less sensitive, were hard to detect and so were about five times as successful at sinking ships as normal noncontact mines. This clearly caused Crick some pride, though mixed with guilt, or at least discretion, in the postwar years.

When in 1943 Massey was spirited away to Berkeley to work on the separation of uranium isotopes, Crick found himself reporting to Edward Collingwood, a Cambridge don who had been invalided out of the Royal Navy after an accident just before the battle of Jutland in 1916 and had then found his métier in academic mathematics. Collingwood appreciated Crick's mind and gave him interesting assignments. They also became friends, and Collingwood later invited Crick to weekends at his grand Northumbrian house, Lilburn Tower. In the winter of 1944–1945, Crick suddenly got an opportunity to travel abroad for the first time in his life. At the time, the newest German weapon was the acoustic torpedo, or Gnat, fired by submarines, which homed in on the sounds of ships' engines. All attempts to recover an intact Gnat had failed. On 30 July 1944 a German U-boat, U-250, attacked and sank a Russian submarine-hunting ship in the Gulf of Bothnia. The explosion brought other Russian ships to the scene, and one succeeded in depth-charging the U-boat, which sank in shallow water. Despite fierce attacks from torpedo boats and shore batteries in Finland, the Russians salvaged the submarine and brought it to the naval base at Kronshtadt with its acoustic torpedoes intact. At first, the Russians refused to share technical details of the capture with their allies, but the British demanded to be allowed to send a team to inspect the Gnats. Eventually, after much delay and argument, in February 1945 Collingwood and Crick were flown via Cairo to Persia so that they could be met by Russian aircraft and flown to Batum and then to Moscow. For the journey Crick was issued a naval uniform with the rank of lieutenant commander in the so-called T-force (for "technical"), of which he kept the cap. He would use it later, when he took up sailing in the Mediterranean in the 1960s.

In Moscow Crick and Collingwood met two Royal Navy officers sent down from the small British garrison at Murmansk, one of whom, Robert Dougall, would become a close friend of Crick's. Dougall recalled in his autobiography his first meeting with Crick: "One was a tall, sandy-haired young man, who walked with a slight stoop. He obviously had an immense sense of fun, which frequently burst out into a high-pitched laugh more like a bray." The party continued to Leningrad, where Crick, with Dougall as his interpreter, spent two weeks in the Peter and Paul Fortress trying to understand the circuits inside the acoustic torpedo. They then returned to Moscow for a further two weeks to compile a report for the Admiralty before Dougall took a train north and Crick a plane south to Persia and back to England.

Chapter Two

Three Friends

THREE PEOPLE ENTERED Francis Crick's life towards the end of the war. All of them would remain part of his life till the end, and all of them would nudge him toward his future greatness. Their names were Georg Kreisel, Odile Speed, and Maurice Wilkins. Kreisel was the first of Crick's intellectual sounding boards. Crick's intellectual technique, throughout his life, was a dyadic pairing, a long-running two-way conversation with a chosen friend, somewhere between an interrogation and a Socratic dialogue. In the periods when he had no such sounding board he was visibly at a loss. Kreisel was the first to take the part, later filled in turn by Jim Watson, Sydney Brenner, and Christof Koch.

Georg Kreisel was seven years younger than Crick but was to be more mentor than disciple. Born in Austria to middle-class Jewish parents, but sent to school in England before the Anschluss, Kreisel went to Trinity College, Cambridge, and became a friend of Wittgenstein. He was a formidable mathematical logician who later made profound contributions to proof theory; and, as befits the species, he was a confirmed eccentric. He generally lived out of a suitcase; went to bed at nine o'clock every night; and to fall asleep, demanded such total silence that he would switch off the refrigerator, and such total darkness that he would pin his own personal blackout curtains over the windows. Except for a few years of cohabitation with Freeman Dyson's wife Verena Huber in the late 1950s, he was a roving bachelor who frequented cathedrals, the beaches of the Riviera, and the castles of the jet set to pursue his conquests. Propositioning women at random on the beach, he claimed a success rate of 10 per cent. Kreisel was a good cook, who in later life would prepare interminable meals at the Cricks' house in Cambridge (he would start cooking each course only when the preceding one had been eaten), stripped to the waist throughout. Once Crick became famous, Kreisel occasionally impersonated him, a fact Crick discovered on receiving a letter from a Spaniard who enclosed a photograph of himself with "Francis"—i.e., Kreisel. The latter was unabashed: "When I travel, I often use your name." On another occasion, when arrested on a Moroccan beach, Kreisel gave the police Crick's name.

Crick had met this unusual character in the cafeteria at West Leigh one evening in 1943. Kreisel had been recruited straight from Trinity to work for Collingwood, though he later moved to London to calculate, among other things, the effect of waves on the "Mulberry" harbours to be used in Normandy. Kreisel and Crick liked each other at once because they both thought that the third person at the table, a chemist, was talking nonsense. Their friendship developed, and Crick later claimed, remarkably, that it was the younger man who taught him to think properly: "When I met him, I was a rather sloppy thinker with a taste for wit and paradoxes in the style of Oscar Wilde. Kreisel would tactfully but sternly rebuke me for any careless thinking so that under his influence my ideas became more logical and better organised." Kreisel believes this refers to the fact that he persuaded the loquacious Crick not to say the first thing that came into his head, but to find a "sharper formulation."

Crick's mind was capable of abstraction—Kreisel once saw him solve the strategy for winning the game of nim from first principles—but it was always to be anchored in empirical facts. By the standards of mathematicians like Kreisel, Crick was a prosaic, even mundane, thinker, but perhaps that was why he achieved so much. Crick had no respect for philosophy, then or later, because he regarded it as a series of disagreements for their own sake, by people who did not trouble themselves with empirical facts and never changed their minds unless you bullied them. But Kreisel was an exception because he was so mathematical in his thinking. Kreisel once refereed an argument between Crick and Wittgenstein in the latter's rooms at Trinity in the spring of 1945. The subject was, of all surprising things, electoral politics. Wittgenstein was complaining about the use of films of concentration camps in Churchill's campaign for that summer's election, fearing it would inure people to the horrors of the camps; and about the difficulty of getting Care packages to his family in Austria. Crick, showing no fear of the great man, dismissed both concerns by arguing that the election would be settled on domestic issues.

Perhaps, too, Kreisel deserves credit for making Crick less

conventional. By all accounts, Crick would emerge from the war very different from the proper young man who had left University College. After the war ended, he was living in a first-floor flat at 56 St. George's Square in Pimlico. By now his marriage to Doreen had unravelled. While still in Havant, she had fallen in love with a Canadian soldier, James Potter, whom she would follow back to Canada and marry. Four-year-old Michael was sent to live with his grandparents in Northampton. After the move to London, Doreen also lived in the flat in Pimlico, in the single bedroom to the right of the entrance hall, while Crick and Kreisel lived in the other two bedrooms to the left. When Kreisel moved out in 1946, he was succeeded by Crick's friend from Russia, Robert Dougall. They employed a Welsh housekeeper to cook their breakfast, and Crick walked or bicycled to work at the Admiralty. Dougall, who had returned to his prewar job at the BBC on leaving the navy and was on his way to fame as the network's chief television newsreader, later recalled that Crick "seemed determined to shake off any stodginess in his makeup." They argued "about religion, politics, world affairs, Russia and so on, and our viewpoint on almost everything was diametrically opposed." (This remark puzzles Kreisel, who cannot recall Crick's getting worked up about any of these topics, except of course religion.) Dougall also believed the atom bomb had a profound effect on Crick, who never wanted to work on weapons again. A picture emerges of a man approaching his thirtieth birthday with little patience for convention, and with a determination to live by his brains and follow his own whims, whatever the world thought. Young Kreisel was partly responsible.

The second person to enter Crick's life towards the end of

the war was the woman who would become his second wife. One evening in early 1945, Crick—who was then still living in Havant—was visiting an office in the Admiralty in London when an attractive young Wren third officer in a smart uniform passed through the room, on her way home from an office upstairs. She dropped her shopping bag, spilling brussels sprouts all over the floor. Crick helped her pick them up and asked her out to dinner. She refused this rather forward proposition from a lanky civilian in an unattractive raincoat, but he sought her out on his next visit to London a few weeks later and asked her to lunch. This time he looked more presentable. "Nothing wrong with lunch," she thought.

Her name was Odile Speed, and she was much that Crick was not: artistic, cosmopolitan, and well travelled. The daughter of a jeweller in King's Lynn and a Frenchwoman who came to Norfolk after World War I to learn English, she had spent two years in Vienna during the 1930s, becoming fluent in German as well as French; and she had been about to go to art school in Paris when World War II broke out. She joined the Women's Royal Naval Service (the Wrens) and after driving trucks for a few months was stationed for three years on the south coast listening to German radio chatter to pass on to the code breakers at Bletchley Park, a job of stultifying tedium. At the end of the war Odile was recruited by the deputy head of the Department of Torpedoes and Mines in the Admiralty, Ashe Lincoln, and given the equally boring job of translating captured German documents relating to torpedoes and mines. Spending long days surrounded by dry, technical engineering tomes, she was desperate to get out of the armed forces, go back to art school, and get on with life. To this end the lanky, ginger-haired man in the raincoat seemed an unpromising means, especially when she learnt that he was married, albeit separated, and had a son. She had no interest in science, and he, at this stage, had very little in art. Yet their lives were to be joined for almost 60 years. In 1945, they began a cautious courtship, under the shadow of his unconsummated divorce, his lack of a career, and the advice of his friends not to add a wife to the burden of looking after his son.

Meanwhile, his career at the Admiralty was no longer satisfying him. There was no doubt that civil servants valued Francis Crick's brain, but they were not so sure about the person who owned it. In March 1946 Crick applied to become a civil servant in order to join the naval intelligence service. Interviewed by a committee of three "provincial professors," he was turned down. Naval intelligence was so eager to have him, though, that it arranged for a second interview—this time chaired by C. P. Snow, the scientist and future novelist. "I did not produce a very good impression," Crick wrote, "but they nevertheless decided to keep me on." He threw himself into bureaucratic turf wars in intelligence, at one point writing to R. V. Jones, former head of scientific intelligence, to ask for detailed help in lobbying key senior officials for a more centralised intelligence office. Was this the moment when Crick acquired his lifelong distaste for administrative work? As an old man, he would say that he avoided all administration because he was no good at manipulating people.

By the middle of 1946 Crick had decided that he was ready to leave government altogether, disillusioned by the bureaucratic muddle then prevailing in intelligence and queasy about using his brain for destructive ends. Later he would look back on this as a moment of decision, not failure. He recalled telling some naval officers about penicillin—not that he knew much about penicillin—and suddenly realising that what he felt like gossiping about was what really interested him: the gossip test, he would call it. But he was 30 years old and had only an unfinished PhD and an aborted career in government science to show for it—not to mention a son to raise. He knew about magnetism and hydrodynamics, both of which now bored him. Most people would have looked for a job in industry or commerce, but inside Crick was still the inquisitive and impatient 10-year-old seized with an ambition to discover something before there was nothing left to discover. He also showed traces of a Kreisel-like bohemian eccentric. He was determined, not just to break into science but to do something heroic in science and, above all, to explode a mystery. He consulted Kreisel, who gave him backhanded encouragement: "I've known a lot of people more stupid than you who've made a success of it." With the bravado of a bankrupt gambler with no high cards left, Crick tried to decide what he would solve first: the secret of the brain or the secret of life. It was the latter that took him to meet Maurice Wilkins, the third significant person to enter Crick's life just after the end of the war.

Wilkins's story was very like his own; and, indeed, since Crick's mother was a Wilkins, they wondered if they might be related. They were not. They were both descended from Unitarian dissenters, though one of Wilkins's dissenting ancestors was famous. They had both been born in 1916 (and they would both die in 2004). They were both born into a precariously middle-class background, but Wilkins's was more intellectual:

his grandmother had attended Cambridge University as one of the first generation of women students. Wilkins's father was an Anglo-Irish doctor who had emigrated from Dublin to New Zealand in 1913 and then returned to Britain in 1923. Wilkins and Crick had both scraped out a disappointing second-class degree in physics (Wilkins at Cambridge), had both started a PhD (though Wilkins had finished his), and had both gone to work on weaponry during the war (Wilkins in the Manhattan Project in Berkeley). They were both divorced, after hasty wartime marriages. By 1946 Wilkins had a promising scientific career as the assistant to John Randall, the newly appointed professor of biophysics at King's College, London. Crick was looking for a job. The smart money for scientific greatness would certainly have been on Wilkins.

Wilkins's goal at this stage was to induce genetic mutations with ultrasound and hope that they would shed light on what genes were. Harrie Massey, Crick's wartime mentor, had set Wilkins on this path while he was in Berkeley by giving him a little book by Erwin Schrödinger called *What Is Life?* This book, a series of lectures delivered by Schrödinger in Dublin in 1943, influenced a whole generation of physicists to go into biology, Crick among them. Read it today, and you will probably wonder what the fuss was about: Schrödinger argued that because genes must be very small, they must be subject to quantum uncertainty, which made it all the more puzzling that they could be stable from one generation to the next. There must be some new physics involved. But then Schrödinger, influenced by his friend Max Delbrück, made one telling remark. Intending to dismiss the idea, he raised the possibility that a gene might be stable if it was an "aperiodic crystal"—that is, something with a regular but not a repeating structure. This was enough to get Wilkins intrigued. (Crick was somewhat less impressed by the book.) Randall recruited Wilkins to Saint Andrews and then asked Wilkins to follow him to King's. Wilkins jumped at the chance—despite the fact that he had already fallen out with Randall twice, once in Birmingham and once in Saint Andrews. This was omen enough of the third and most fateful misunderstanding, still to come, over Rosalind Franklin's appointment in 1950.

So Wilkins was already embarked on the search for the secret of life when Crick met him in 1946. Crick went to see him at Massey's suggestion. Wilkins liked Crick and wanted Randall to hire him; but Randall thought he talked too much, and the idea came to nothing. Crick was in any case unimpressed by the research at King's: the researchers seemed more interested in their instruments than their samples. Crick later said that Wilkins was wasting his time trying to study DNA and told him to "get himself a good protein." Wilkins and Crick became friends, though, and Wilkins came to supper in Odile's flat in Hogarth Road, where he made himself memorable by going straight to the kitchen to peer at what was cooking.

By now, Crick was at least sure he wanted to solve the problem of life. The brain would be fun, too; but, inspired by Schrödinger, he thought a physicist could do more about life. So he had already turned down an actual offer of a research position working on colour vision. Once he had decided what he wanted to do, he applied to the Medical Research Council (MRC) for a studentship, explaining in his application: "The particular field which excites my interest is the division between the living and the non-living, as typified by, say, proteins, viruses, bacteria and the structure of chromosomes."

Perhaps the most extraordinary fact about Francis Crick at this stage in his life was that he had comprehensively reeducated himself in his spare time. Ever since the last years of the war, he had been reading everything he could lay his hands on in physics, chemistry, and biology. He got leave from the Admiralty during working hours to attend seminars in theoretical physics. Sitting at his desk, he would surreptitiously read a book—so would many people, but not a textbook on organic chemistry. In July 1946 he read an article in *Chemical and Engineering News* by a man with an unusual name who argued that biology would be explained not by strong intramolecular forces but instead by the newly discovered weak attraction between two molecules, one of which had a hydrogen atom attached: the hydrogen bond. Crick did not realise that Linus Pauling was the most famous chemist in the world, but he stored away the idea. Nor was it just books that Crick studied. Michael recalls him bringing frogs to his parents' home in Northampton on weekends, to dissect them on the steel table kept in the house for shelter during air raids. Crick almost never read a newspaper, for two eminently rational reasons: first, because if anything really important happened he would hear about it from people in the street on his way to work; second, because working in intelligence had persuaded him that the real stories never reached the newspapers. He read science.

The border between the living and the nonliving was a frontier that had been shifting since at least 1828, when Friedrich Wöhler synthesised urea, a chemical hitherto found only in living bodies. The search for the vital spark that made flesh so utterly different from clay had gradually become a matter of genetics. By the early twentieth century nearly everybody except diehard vitalists thought that the main thing which made living things different was not that they had some special goo called protoplasm operating a different chemistry set, but that their genes mysteriously enabled them to make copies of themselves. In 1865 in Brünn (Brno), Gregor Mendel had realised that he could explain his plant breeding experiments only by postulating that inheritance came in discrete, concrete "factors"—they acquired the name "genes" later, in 1909. Thereafter Thomas Morgan proved that genes were linked in linear sequences, Theodore Boveri that they resided on chromosomes, Hermann Muller that genes were mutable by X rays, and George Beadle that each different chemical reaction in a cell was effected by the product of a different gene.

The concept of the gene was therefore already central by the middle of the twentieth century. But it was an entirely abstract concept. The truth was that nobody had the faintest idea what a gene actually was. It might give you blue eyes or brown, but by what means? You can trawl back through the effusions of the era and find the occasional prophetic remark. J. B. S. Haldane spoke in 1934 of two-dimensional genes copying themselves by means of "negative" templates, the first hint of complementarity. Dorothy Wrinch suggested, also in 1934, that the fact that genes were linked to each other in linear sequences and so were amino acids in proteins might be more than a coincidence: the first hint of a coded sequence. But nobody followed up either idea, and both were lost in a crowd of mistaken ones. As late as 1950, in an essay to mark the fiftieth anniversary of the rediscovery of Mendel,

Hermann Muller was able to state: "We have as yet no actual knowledge of the mechanism underlying that unique property which makes a gene a gene—its ability to cause the synthesis of another structure like itself." When *Life* magazine published, in 1949, a hugely magnified photograph of part of a chromosome, claiming that this was the first photograph of a gene, it merely emphasised this ignorance: how would you recognise a gene even if you saw one?

Chapter Three

Cambridge

CRICK'S SEARCH for a post in science did not go smoothly. An application to join the famous J. D. Bernal's crystallography lab at Birkbeck College was tartly rebuffed by a secretary: did he not realise that everybody wanted to work with Bernal? Several other introductions led nowhere. His wartime reputation was high enough that the Medical Research Council (MRC) treated him as a prize catch for biology, but it could not, at first, present him with a place. Sir Edward Mellanby, secretary of the MRC, was embarrassed to offer a "man of his standing" a mere studentship, at £350 a year, but Crick was not worried about money, and his application was approved even though he still had no berth in a laboratory. Eventually, after several other options fell through, Mellanby sent Crick to see Honor Fell, the director of Strangeways Laboratory in Cambridge, who agreed to take him on because her resident physicist had just died. Crick gave his notice to the Admiralty and moved to Cambridge in September 1947. He took lodgings in Jesus Lane, though he still spent every weekend either in London with Odile or in Northampton with his parents and Michael.

Strangeways was a biological laboratory founded as a charity by Dr. Thomas Strangeways in 1905. It was housed in a large, handsome redbrick villa on the southern edge of Cambridge, a few miles from the city centre. Though loosely affiliated with the university, it was a private institution where Honor Fell was perfecting the art of growing human cells in an artificial culture. Crick joined the lab of Arthur Hughes, who had ingeniously managed to persuade cells to ingest tiny magnetic particles and then, by subjecting the cells to a magnetic field, could make the particles move through the cell. Hughes needed an expert on viscosity and magnetism to tell him what this procedure revealed about the properties of the cell's insides.

That winter Crick received a letter from Wilkins:

> My Dear Crick,
>
> How is Cambridge? Is the cold wind blowing across the fens, frisking up the waters of the Cam, whistling through the barbed wire on college walls, rattling the chain padlocks on college gates and causing a healthy glow to appear in the faces of bedmakers and undergraduates hurrying across the cobbles to the college bathroom? Is it blowing in under the door of the Strangeways, congealing the culture media and causing all honest amphibians to hibernate? In fact, how are you getting on and—by the way:—(no that is not to be mentioned). And when you come to town next send me a card in advance and I can

reply by phone and suggest a date for dinner. I have made some very good dinners lately and am getting in a barrel of cider.

Do let me know, won't you?

Yours

Maurice Wilkins.

Crick later saw Strangeways as an apprenticeship in biology, equipping him for the big questions that were to come. And the MRC was happy to have found somewhere to park him for his studentship. But at the time, Strangeways must have seemed like a dead end. Not only was it a long way from the middle of Cambridge (Crick had not yet learnt to drive, let alone acquired a car), but, being independent, it could not register him for a Cambridge PhD. Besides, he was back to measuring viscosity. In January 1948 his father died, at age 60. Wilkins wrote that he hoped this would not upset Crick's finances—but rich Uncle Arthur's support was still there.

Putting a brave face on the situation, Crick threw himself into the experiments. He joined the Natural Sciences Club and made friends with the zoologists Michael Swann and Murdoch Mitchison, whose polarising microscope he used. With Hughes, he also travelled to Paris to see the expert on lipids, Dikran Dervichian, at the Pasteur Institute. Eventually, Hughes and Crick published two very long, detailed papers: one full of equations; the other full of experimental details about how to "twist," "drag," and "prod" magnetic particles through cytoplasm. The papers reach no definite conclusion—sometimes particles rebound as if stuck in a gel; sometimes they do not—and are full of statements like, "It can be seen from the figures that the results are by no means clear-cut." This is, frankly, science at its worst: more than 70 dense pages of detail-obsessed, overanalysed measurement for its own sake, with no hypothesis. It is the kind of thing with which the journals are stuffed, and which nobody ever reads. Yet even here there are hints of Crick's future style. The long-forgotten grand panjandrums of cytoplasmic studies come in for sharp, slightly patronising criticism: "Heilbronn and Heilbrunn [sic] . . . do not seem to have realised that in all probability such forces were entirely swamping any effect due to viscosity." "We particularly deprecate statements such as that of Frey-Wyssling." And an especially Crickian remark: "This may be true, but there does not appear to be any evidence for it."

Once during this period, Crick gave some visitors to Strangeways a seminar on the problems of biology at the molecular level. Though in later years he could not recall precisely what he said, he knew he had mentioned one fact about DNA—that its viscosity was much reduced by X rays (implying that X rays fragmented large DNA molecules). He thought he had probably said more about DNA, including the theory that genes are made of it. Certainly, sometime between 1946 and 1951 he seems to have switched from believing that genes were made of protein to believing that they were made at least partly of DNA. In this, he was not the first, but not the last either.

DNA had a history very like that of the gene itself: both were orphans of their science. Just as Mendel's insight of 1865 was forgotten for 35 years, so too was "nuclein" neglected by geneticists after its discovery in 1868 in Tübingen by Friedrich Miescher. He purified a phosphorus-rich, acidic substance from the pus-soaked bandages of wounded soldiers, calling it nuclein because it seemed to be richly present in cell nuclei. Later, in Basel, he got purer samples from salmon eggs. Nuclein was renamed desoxyribose nucleic acid and later deoxyribonucleic acid, or DNA; and for most of the first half of the twentieth century it was thought to be a sort of scaffolding on which genes rested. DNA obviously came in fairly large molecules, and those molecules obviously had a monotonous structure: a simple phosphate joined to a pentagonal sugar ring to another phosphate to another sugar and so on. The only break in the monotony was that also attached to each sugar was an organic nitrogenous "base," either a ring or a double ring of carbon and nitrogen atoms, and these bases came in four different kinds: adenine, guanine, cytosine, and thymine. This was surely not enough variety to explain the complexity of life.

Yet since the mid-1930s, Oswald Avery at the Rockefeller Institute in New York had been steadily and painstakingly accumulating evidence that—at least in one highly specific case—purified DNA seemed to have the properties of a gene: it could change the nature of a creature in a way that was heritable. Avery published his experiments in a long paper in 1944, describing in detail how a pneumococcus bacterium of one nonvirulent strain could be turned into another, virulent strain simply by mixing it with purified DNA from a virulent strain. Avery went to great lengths to ensure that his extract was pure, repeatedly rinsing it in chloroform, enzymes, and alcohol to remove all protein, till he had left just one-hundredth of an ounce of material extracted from 20 gallons of bacterial broth. Test after test confirmed that the "transforming substance" had all the properties of DNA and none of the properties of protein.

Yet Avery did not persuade the world. It was not that his paper was too obscure, either in language or in distribution. Almost everybody in biochemistry and genetics knew of the experiment. What happened was a classic case of psychological vested interest of the kind that science knows only too well. People had invested heavily in protein genes. For years, they had absorbed Phoebus Levene's orthodoxy that DNA was a "stupid" repetitive substance with no variation: it could not have the specificity of a gene. The Rockefeller Institute itself had recently humiliated a German scientist who claimed that some enzymes were not made of protein, so it was steeped in protein partisanship. And Avery's colleague at Rockefeller, Alfred Mirsky, kept up a sustained and somewhat personal campaign in favour of the theory that Avery's result was explained by faint protein contaminants in the samples. So Avery could not even command the support of his own institution. Besides, many people doubted that bacteria had genes at all, or if they did have genes, that those genes were made of the same stuff as in animals. All in all, by the late 1940s, some people thought genes were made of DNA, some people still thought they were made of protein, some people thought they were a mixture of the two, and some people were undecided. Crick was probably in the last category, but he was leaning towards DNA. In any case, though, he was just a distant spectator of the debate.

For the moment it was proteins, and Kreisel, that got him out of Strangeways. Possibly at Crick's request, Kreisel went to see a fellow Austrian, Max Perutz, who had recently been appointed head of the newly constituted Medical Research Council Unit for Research on the Molecular Structure of Biological

Systems at the Cavendish Laboratory, and asked him if he would take Crick on. Perutz was keen, so Crick went to see him, and Mellanby happily agreed to transfer Crick's studentship to the newly funded group. In the summer of 1949, Crick shook the dust of Strangeways off his shoes and headed towards the heart of Cambridge, the heart of the university, and the heart of science—though he was taken aback when he first asked a taxi driver at Cambridge station to take him to the Cavendish and the man had never heard of it. The Cavendish was Britain's most famous physics laboratory, the home of James Clerk Maxwell, J. J. Thomson, and Ernest Rutherford.

Perutz had come to Cambridge quite voluntarily in 1936 to work with J. D. Bernal, but the Nazis had made him an impoverished exile by confiscating his family business and driving his family out. The British had then interned him as an enemy alien during the war, first on the Isle of Man and then in Canada; these grim experiences would have embittered a less gentle man. His job, working under the head of the Cavendish, Sir Lawrence Bragg, was to do for biological substances what Bragg had done for common salt in the early years of the twentieth century: discover their structure using X rays.

In 1912 Max von Laue and his colleagues had shown that crystals of copper sulphate diffracted X rays, thus proving that X rays were waves. But it was young Lawrence Bragg, still a student at Cambridge, who saw that the diffraction pattern must contain clues to the structure of the crystal. With his father, Sir William Bragg of Leeds University, he worked out the detailed mathematics necessary to recover the structure of the crystal from the pattern of spots left by the diffracted rays. Lawrence

Bragg was on active service in France in 1915 when he learnt he and his father had won the Nobel Prize.

Since then crystallography had proved its worth with ever more complex crystals. Two generations of the Braggs' students had spread it across the country and had begun to attack biological molecules. J. D. Bernal in Cambridge and William Astbury in Leeds had first shown that proteins could have sufficiently crystalline order to give good X-ray diffraction images—Bernal with pepsin and Astbury with keratin. One of Bernal's students, Dorothy Crowfoot (later Hodgkin), then X-rayed insulin crystals in Oxford, while another, Perutz, embarked on the much larger molecule of haemoglobin in Cambridge and proved that it had a definite, nearly spherical structure and was not an amorphous colloid. There were intriguing hints that Perutz and Astbury were picking up similar patterns despite their very different proteins—one a fibre, the other globular. It seemed only a matter of time before one of these four friends would unveil the natural structure of protein. But the war came and went, research gradually resumed, and still no breakthrough came.

In 1949 Perutz's team consisted of John Kendrew, a chemist who had been a scientific aide to Lord Mountbatten in the war; and his student Hugh Huxley (with Lawrence Bragg looking down from above and Tony Broad, an electrical engineer, to help them with the unusually powerful rotating-anode X-ray machine). Crick became the fourth scientist on the team and Perutz's first student, even though he was just two years younger than his supervisor. His job at the Cavendish was no better paid than the job at Strangeways—he was effectively just transferring his studentship to another institution—but, with the divorce from Doreen having come through in 1947, Crick now felt secure enough to propose marriage to Odile. She decided to give up her course in fashion design and move to Cambridge.

They were married on 14 August 1949, Odile wearing a knee-length dress she had designed herself, Francis in a morning coat. After the wedding, in a registry office, the reception was in the garden of a house in Cheyne Row in Chelsea. They left by train for their honeymoon in Liguria in northern Italy, where they sought out a remote little hotel in Punta Chiappa, accessible only by sea, that Crick had heard about from an old school friend. The bridal room gave a view straight out to sea over the cliffs. It was a pleasant respite from Austerity Britain. They then returned to Cambridge, to a little flat above a tobacconist's shop on Thompson's Lane, opposite St. John's College, which had recently been vacated by the Perutzes. It was called the Green Door and, though comfortable enough, was far from luxurious. The bath was in the kitchen, hidden under a folding board that was usually covered with dishes. The lavatory was halfway up the stairs; here Crick shaved every morning in the small washbasin—thinking hard while shaving would be a lifelong habit. The flat had a bedroom, a sitting room, and a small room for Michael when he came home from boarding school at Dunhurst. The rent was 30 shillings a week. Money was tight. Several times the Cricks had to take their typewriter to a pawnshop on Bridge Street in order to raise a little cash.

Crick was now making his third attempt to get a PhD. He could not afford to fail again. But the problem he was to attack—essentially to choose a protein and discover its structure—had defeated Perutz for more than a decade, for the seemingly insuperable reason that an X-ray diffraction pattern records only the intensity of the waves, not the relative timing when each wave arrives at the plane of the picture. This so-called "phase problem" (the name deriving from Fourier analysis) could be circumvented in the case of small molecules by trial and error with model-building, as Lawrence Bragg had shown many years before. Crick put it thus: "If the structure could be guessed, it was merely a problem in computation to derive the X-ray pattern it should give. This put a high premium on a successful guess." But for giant globins, there were too many options for such guesswork to yield results yet. A smaller protein might be easier.

Crick first took up secretin, a tiny hormone found in the intestinal tract, but he could not crystallise it easily. He had more luck with a comparatively small protein, trypsin inhibitor, which he was able to crystallise by evaporating a solution of it very slowly for several weeks, in a flat-bottomed jar with a capillary through the cork stopper. The crystals, a few tenths of a millimetre in length, proved to have a disappointingly large unit cell (or minimum crystal size), containing about 60 molecules, so an X ray would tell him little about the individual molecule. He next tried lysozyme, an antibacterial protein found in human tears and birds' eggs, which crystallised easily and proved to have a smaller unit cell. He tried lysozyme from different species of bird, looking for one in which the protein crystallised differently. But the results were no better in the eggs of guinea fowl, turkeys, ducks, or geese. In the eggs of lesser black-backed gulls he could find no lysozyme at all. Eventually he drifted back to helping Perutz with haemoglobin. Prompted by Lawrence

Bragg, they now saw an advantage in analysing the same kind of protein from different species, so Crick's early notebooks are full of references to ox, horse, and rabbit haemoglobin.

When Crick arrived, Perutz had just published a paper about haemoglobin, hopefully offering a four-layer structure known as the hatbox model. Crick's first act was to point out the flaws in this. He had been reading for several months, and his own interpretation of the densities implied by the X-ray data showed a much more haphazard and less regular structure than Perutz had envisaged. No more than one-third of the protein could consist of parallel chains of polypeptides. In effect, Crick was killing off the notion that proteins, though regular, had simple geometric structures—and with it the hope that protein structure would quickly reveal the secret of life. Crick had characteristically found his own, primarily visual way to understand protein crystallography. This was to be his unique contribution to the subject. It was not that he avoided the immensely laborious calculations involved in deducing structures from images—the Fourier analysis, Bessel functions, and Patterson calculations—but that he could intuitively see, in his head, the space-group symmetry of a crystal's unit cell: how you would rotate it to make it look the same again. He worked hard at this, squinting in a special way at models to see them stereoscopically. In years to come Crick would talk easily in terms of symmetry while others struggled to visualise what he meant. "Although it is necessary to be able to handle the algebraic details, I soon found I could see the answer to many of these mathematical problems by a combination of imagery and logic, without first having to slog through the mathematics."

Crick gave his first seminar in 1950, a 20-minute talk on the theory of protein crystallography. It was almost entirely negative. The title of this talk (after a line in Keats's "Ode on a Grecian Urn") was "What Mad Pursuit?" Crick worked his way through all the methods being applied by Perutz and others and demonstrated ruthlessly that they were doomed to fail—all except one: isomorphous replacement of heavy atoms with other elements. He was absolutely right, though it was only Bernal who later acknowledged Crick's role in steering protein crystallographers in this direction. Later, Perutz and Bragg would finally crack the phase problem for haemoglobin with the help of multiple isomorphous replacements. But at the time, Crick's assertion was, to say the least, tactless. As usual, Crick gave no thought to softening the blow. He actually made it worse some time later, over tea, by treating Bragg to one or two condescending asides implying that Bragg—who hade invented crystallography—might not know much about the subject. At one meeting, Crick was chatting in his usual critical vein within earshot of Bragg, who exploded: "Crick, you're rocking the boat."

Bragg and Perutz needed the patience of saints to put up with this loudmouth with the braying laugh who was much better at telling them what was wrong with their science than actually making measurements himself. Perutz had such patience; Bragg did not. What made it worse was that Bragg was about to be humiliated by an old rival, Linus Pauling. Bragg, Perutz, and Kendrew had begun another approach to protein structure by trying to devise a plausible structure for typical polypeptide chains, rather than whole protein molecules. Astbury in Leeds had produced X-ray evidence that long stretches of the polypeptide chain in keratin, the protein of which wool is made, had a natural repetitive shape. The three scientists at Cambridge decided to build scale models using metal bonds and atoms to see if they could guess the structure of the chain. Obviously, this might simply be a straight line, but that would be highly unlikely. Much more plausible was a helix of some sort in which the angle connecting each amino acid to the next gave the structure a natural twist. They were misled by a crucial spot in Astbury's X-ray patterns, one which implied some kind of repeat every 5.1 angstroms (an angstrom is one 10-billionth of a metre). Assuming that this was caused by the "pitch" of the helix—the distance between one twist and the next—they concluded that there must be a whole number of amino acids per turn, probably four. This gave unsatisfactory-looking helices with (as it turned out, and as every biochemist knew, if only any biochemists had been asked) an impermissible angle to the peptide bond. But they published it anyway.

Shortly afterwards, Pauling published a much more elegant structure for keratin, which he called the alpha helix. Despite the fact that his structure had only 3.6 amino acids per turn and could not explain the 5.1-angstrom spot, it was soon apparent that Pauling was right and Bragg was wrong. Indeed Perutz proved as much almost immediately by finding a 1.5-angstrom spot on the vertical meridian (caused by the "rise" from one amino acid to the next) that Astbury had overlooked.

At this moment, when Bragg was touchy about his humiliation by an old rival, Crick, not content with rocking the boat and patronising the great man, tactlessly told him that one of his newest ideas was old hat. In October 1951 Bragg circulated a draft of a paper containing the "minimum wavelength principle," an especially ingenious use of Fourier analysis—and an idea that Crick claimed to have had nine months before. This was the last straw. Bragg made it clear that he was insulted by Crick's insinuation. He fired off a furious letter to the Medical Research Council about Crick, summoned Crick to his office, and told him he had no future at the Cavendish after finishing his PhD. Crick was visibly shaken.

But life was about to get better. On 31 October 1951 Bragg showed Crick a paper he had just been sent by Vladimir Vand in Glasgow, claiming to derive the general pattern that would always betray the presence of a helix in an X-ray diffraction. Crick consulted another physicist, Bill Cochran. Both Crick and Cochran noticed that Vand's paper was only half right. After lunch Crick went home to nurse a headache and there, at the Green Door, sitting in front of a gas fire, he puzzled out the correct solution before recovering enough to go to an evening tasting at Matthew's wine merchants in Trinity Street—an event he had been eagerly looking forward to. Nineteen different hocks and Moselles of the 1949 vintage were on offer; Crick tasted all of them, noting his reaction to each carefully on the sheet of paper, but not spitting out the mouthfuls as was customary. The headache presumably returned.

The next morning he found that Cochran had also derived the identical formula, though with a less clumsy proof. Essentially all they had done was unknowingly repeat what Alexander Stokes had managed at King's in London a few months before, but for Crick it was a euphoric moment. He had actually made a discovery, found a general law about nature, made a positive contribution to others' labours. Bragg was slightly mollified: it was a smart piece of mathematics. While it did not solve the phase problem, it did make it possible to predict the diffraction pattern that a helix of given dimensions would produce.

A year later, Crick just managed to beat Pauling to publishing an explanation of the puzzling 5.1-angstrom spot. It was caused by a "coiled coil." Because there was not a whole number of amino acids per turn, alpha helices could not stack together neatly but had to coil around each other, slightly deformed, the protruding amino acids from one helix fitting like knobs into holes on the other. This distortion produced the repeat at 5.1 angstroms.

The upshot of Crick's first two years at the Cavendish was therefore that he was now acknowledged, even by Bragg, as a good theoretician, though he was correspondingly not of much use in the lab and was not apt to see things through. He was an irritating presence, because of what Bragg called his habit of "doing other people's crosswords" as well as freely criticising their best ideas; but he had learnt some useful lessons about the need for simplifying assumptions and the importance of visualising reality as well as analysing it—not to mention the importance of never being beaten by Pauling again. All these ingredients would be crucial in the story of the double helix. However, Crick was no closer to the definition of life. Fascinating as the arguments over protein crystal symmetry were to him—they were better than viscosity, at any rate—they did not promise an immediate insight into life's mechanism. And he was going to be without a job when he finished his thesis, because Bragg did not want him around.

Chapter Four

Watson

JAMES WATSON ARRIVED in Cambridge in September 1951. The first Crick he encountered was Odile, to whom Perutz introduced him as they met in the street. She was pushing a high pram with the new baby, Gabrielle, in it. Legend has it that she told Francis later, "Max was here with an American who had no hair." (Watson had a crew cut.) It was three weeks later that Watson actually started work at the Cavendish and met Francis Crick for the first time. Watson describes it as an instantaneous meeting of minds. Within half an hour they were talking about guessing the structure of DNA.

This was Watson's obsession. He was tearing round the world, impatiently looking for somebody who could help him find the structure of the gene. Inspired by Schrödinger's book, he expected genes to be molecules; convinced by Avery's experiments, he knew that they were made of DNA. He had experienced these epiphanies before he graduated from the University of Chicago.

Watson was the son of a professional bill collector and amateur ornithologist from the south side of Chicago. He had gone to the university at age 15, had graduated at age 19, and had received a PhD from Indiana University in Bloomington a month after his twenty-second birthday. He had gone to Bloomington hoping to work with Hermann Muller, the man who had first mutated genes in fruit flies artificially. Actually, though, Watson was more attracted by the genetics of Salvador Luria involving the "bacteriophage" virus. Watson struck Luria as "odd." He was tall, stick-thin, socially ill at ease, and apt to snort with laughter halfway through a sentence. He was also in the habit of speaking his mind with startling frankness. During graduate school Watson came to know and venerate the hero of Schrödinger's book, Max Delbrück, spending two summers with Delbrück at Cold Spring Harbor and another at Caltech.

The "phage" world, excellent for studying how viruses mutated, duplicated, or recombined, still offered no insight into what a gene was. Delbrück and Luria were not especially interested. So Watson went to Copenhagen to work with Herman Kalckar, who was studying nucleic acids. But Copenhagen proved a dead end, too. It was dark and wet, and Kalckar was interested only in chemistry, not structure—when he was not preoccupied with his own divorce. Watson did some experiments in a different lab with Ole Maaloe, using radioactive phosphorus, then travelled with Kalckar to Naples to the famous Stazione Zoologica di Napoli in the spring of 1951. While he was there, suffering the unexpected cold of an Italian spring, there was an international meeting on macromolecules to which Maurice Wilkins came as a late substitute for John Randall. At that meeting, suddenly, Watson saw Wilkins show an X-ray photograph of DNA.

Wilkins had been taking X-ray photographs of DNA for almost a year at this point. In May 1950, at a meeting in London, he had been given an unusually intact sample of DNA prepared from calf thymus glands by the Swiss biochemist Rudolf Signer. He found that he could draw it into thin fibres, which looked remarkably uniform under the microscope; tried X-raying the fibres as if they were crystals; and went to Raymond Gosling, a graduate student in the department, and asked him to help. Gosling set up Randall's old, rather feeble Stubbins X-ray gun in a lead-lined basement. At first he could get nothing, because the air scattered the X rays. He tried pumping hydrogen into the apparatus, and—worried about the risk of an explosion—decided to measure the flow of hydrogen by bubbling it through water. This fortuitously added humidity to the hydrogen, a crucial if serendipitous step. With some plasticine and a condom (from Wilkins's wallet) judiciously placed to prevent the gas from leaking, the improvised apparatus at last worked. Each exposure took 20 or 30 hours but produced a startlingly clear pattern of simple spots, far simpler than what proteins produced, or than what Astbury had managed to generate by X-raying DNA. It was these pictures, taken the previous summer, that Watson saw in Naples.

Watson was electrified: he saw immediately what Wilkins had already concluded: that genes must have a regular, symmetrical structure. This was a genuine surprise, given how much DNA varied, from species to species, in the proportions of its nitrogenous bases. How could it be both regular and variable? Using his pretty sister Elizabeth as conversational bait, Watson tried to talk to Wilkins during an outing to the ruins at Paestum, hoping to persuade Wilkins to offer him a job. Failing in that—Wilkins found his conversation hard to follow—Watson determined to get to another X-ray crystallography laboratory. The long-suffering Luria eventually asked John Kendrew to take Watson in at the Cavendish.

Whomever Watson went to work with, he always switched to collaborating with somebody else. In Bloomington, he had abandoned Muller for Luria for Delbrück; in Copenhagen he abandoned Kalckar for Maaloe. And at Cambridge he soon tired of the caution Kendrew and Perutz expressed about whether genes were made of DNA. "Too many people want things 99 per cent proven before they act on it." Then he met Crick. It is not clear if Crick was already convinced that genes were made of DNA or if Watson convinced him; but either way, Watson found that Crick "did not need much persuasion." Crick later wrote that he had been asking himself the question "where proteins came from" since before he went to Strangeways. It had not, however, occurred to him until Watson arrived that he might find the structure of DNA. Within weeks, Watson was writing to Delbrück that Crick "is no doubt the brightest person I have ever known and the nearest approach to Pauling. . . . He never stops talking or thinking." Crick was thrilled to meet somebody who knew genetics and geneticists. He and Watson started teaching each other what they knew. Explaining Fourier analysis and Bessel functions to Watson was harder than explaining phage mutations to Crick. At one point Crick joked about writing a paper called "Fourier Transforms for Birdwatchers" especially for Jim.

Soon they were having lunch together almost every day in the Eagle, a rambling pub that belonged to Corpus Christi College, in Benet Street just off Kings Parade, about 100 yards from the Cavendish. They usually ate in a room at the back known as the RAF bar. It had been popular during the war with British and American airmen, and its ceiling was covered with graffiti made by the pilots' cigarette lighters—smoky squadron numbers and slogans. "Ethel of the Eagle," a large lipstick drawing of a woman wearing nothing but a cigarette, looked down on their meals. Afterwards, they would often stroll through the grounds of King's College and walk along "the Backs" by the River Cam deep in conversation. With an occasional summer punting expedition, and with frequent breaks for coffee at ten-thirty in the morning and tea at four in afternoon, talk replaced work, or work became talk. Sometimes, Crick would take Watson home for a good meal cooked by Odile; sometimes Watson—appalled by the dreary food offered at Clare College—would simply turn up looking hungry at dinnertime.

A shared passion for scientific gossip was the essence of this dyad. Because each told the other when he was talking nonsense, yet neither felt the least inhibition about sharing speculative thoughts, they could explore the ocean of the unknown without ever getting too far from the coast of facts. "We weren't the least afraid of being very candid to each other, to the point of being rude," said Crick many years later. In its formal forums, science discourages speculation, and too few scientists make room for it even in informal settings. Both Crick and Watson later recognised that there was something fraternal in their friendship, with Watson eager to play the role of the younger brother—admiring, but also competitive.

Maddened by the incessant chatter, Perutz and Kendrew

put Crick and Watson together in a room that had just become available down the corridor from their laboratory, on the first floor of the Austin wing of the Cavendish, a functional four-storey rectangle of brick opened by the car manufacturer Lord Austin in 1939. It was a tall-ceilinged room, about 20 by 18 feet with a 13-foot ceiling. Today it is largely unchanged, its walls still lined with whitewashed bricks over which run a few broad wooden laths to one of which the first diagram of DNA was pinned. Two large metal-framed windows look east into a clutter of other buildings. At first Crick and Watson had the room to themselves, but soon there would be new arrivals.

There was only one problem: neither Crick nor Watson was being paid to study DNA. Crick's subject was supposed to be haemoglobin; Watson's myoglobin. True, after some disastrous laboratory experiences—Crick had twice flooded a lab by mis-attaching a rubber tube to a suction pump—neither of their supervisors really missed their presence. But they were playing truant nonetheless, and they had no experimental data to go on. The best data were all at King's College, London. So, prodded by Watson, Crick now invited Maurice Wilkins up to Cambridge for a weekend, hoping to hear more.

Wilkins had not been idle since his trip to Naples. He had gone there to collect sperm from cuttlefish so as to take X rays of the heads of the sperm—filled with DNA—and confirm that the pattern produced by the DNA Signer had prepared was true of DNA everywhere. The pattern was still there, and in herring sperm, too. All DNA, when X-rayed, produced a pattern of spots with no spots in the "meridian" above and below the centre of the image. According to the physicist Alexander

Stokes, of King's, this fact implied a helix of some sort. (Seen from the side, a helix is effectively a zigzag structure; the zigs throw the X rays one way and the zags throw them the other way, leaving a gap in between.) Wilkins had showed the pictures at a colloquium, organised by Perutz, in Cambridge in July and had repeated his argument for a universal DNA structure that included a helix. Wilkins even suggested that the angle of ascent of the helix must be 45 degrees, its diameter 20 angstroms, and the "height" of each turn 27 angstroms. Crick was there, sitting in the back row, but this was three months before his own epiphany about the helix, when he had the headache at the Green Door, and he did not even remember Wilkins's talk. That is a measure of his lack of interest in DNA in the summer of 1951, before Watson arrived.

Ironically, within minutes of his talk in Cambridge in July, Wilkins's triumph had turned to ashes. On the way out he was greeted by his comparatively new colleague, Rosalind Franklin, who told him quietly and firmly to stop working on DNA: "Go back to your microscopes," she finished. Franklin was under the impression that the DNA work at King's was now hers, for the simple reason that Randall had told her so. Randall wanted Wilkins to give it up, and the previous December he had recruited Franklin, an accomplished X-ray experimentalist, to take it over—only he had neglected to tell Wilkins this. Wilkins thought Franklin had been recruited to help him—indeed, he thought he had suggested it.

Rosalind Franklin was a physical chemist who had been educated at Cambridge. She came from a wealthy and prominent Jewish family: her great-uncle Lord Samuel was a former home secretary who had written the memorandum that led to the Balfour Declaration and hence to the creation of Israel. Her grandfather was senior partner of Keyser's bank. After Cambridge she had gone to Paris to study the structures of coal, graphite, and other forms of carbon; in Paris, she gained a reputation for quick wit and skilful use of X rays. Recruited by Randall from the heady atmosphere of bohemian Paris to stuffy, hierarchical King's, she became unhappy soon after she arrived, not least because of the strange attitude of the semi-silent Wilkins, whose meandering conversation (usually delivered at an angle of 90 degrees from the listener) never seemed to get to the point, and who seemed reluctant to hand DNA over to her. She was also quite new to DNA, which she approached as a chemical, not a biological, puzzle.

After the confrontation with Wilkins in July, Franklin had spent the summer rebuilding the X-ray apparatus and, with Gosling's help, started taking pictures of Wilkins's "Signer DNA." She was annoyed that Wilkins was still working on "her" project. He was surprised that she did not want to continue what he had expected to be a collaboration. Wilkins with his hesitant shyness and she with her prickly abruptness brought out the worst in each other. So during that October weekend in Cambridge, Wilkins told Crick and Watson that he was as much in the dark as they were about what she had found. He hoped to learn more at a colloquium they had organised on 21 November.

Watson immediately asked Wilkins if he, too, could attend the meeting. For Crick to go would seem more threatening; and besides, Crick was not yet treating DNA as more than a side interest. Wilkins agreed. A few weeks later, therefore, Watson took the train to London, where he watched Franklin show new photographs, which she had taken that autumn using better apparatus and more carefully prepared samples. She had used saturated salt solutions to keep the humidity constant within the apparatus, and this had enabled her to photograph DNA in a moister, "paracrystalline" state, known from then on as the "B form." Her notes reveal that she too thought at least one form was helical: "suggests a spiral structure."

At the colloquium she mentioned a crucial fact about the drier, crystalline A form. From the X ray she had been able to calculate what the "space group" of the crystal was—in other words what kind of rotational symmetry it had. It was C2, or "face-centred monoclinic," according to a classification of 230 different kinds of crystals drawn up in the nineteenth century. If Crick had heard this, it would have changed history, or so he later claimed. A monoclinic object has a twofold axis of symmetry—it must be rotated through only 180 degrees before it looks the same again. Two pencils taped together facing different ways form a monoclinic object; two pencils taped together facing the same way do not. The face-centred part would have told Crick that the axis of symmetry was perpendicular, not parallel, to the fibre axis. Because Franklin also gave sufficient dimensions to predict that the symmetry axis went through a single molecule, rather than through a crystal made of several molecules, Crick would have built a two-strand helix model with the chains running in opposite directions. He was probably the one person who would have seen this immediately. His trypsin inhibitor, when bound to trypsin itself, crystallised in a face-centred monoclinic way, as did ox haemoglobin. That autumn Crick was living and breathing self-taught helical diffraction theory.

Yet Crick was not at the colloquium. The very next morning he took a train from Cambridge to London on his way to Oxford to make the most of his breakthrough helical theory by telling Dorothy Hodgkin about it. Watson joined him at Paddington Station, and on the train Crick began to interrogate Watson about the colloquium at King's and the new pictures shown by Rosalind Franklin. Watson had taken no notes. He had been learning crystallography for little more than a month. So he misremembered several key facts, notably the quantity of water in the fibres; and he did not mention the space group. But he did at least remember some of the key dimensions. Crick began to doodle on the back of a paper. By the time they were near Oxford, he had decided that there were only a few arrangements that would fit both Rosalind Franklin's X rays and his own helical theory. They should, he thought, imitate Pauling and build a model. According to Watson, Crick spent the day in Oxford telling everybody that they had an idea for the structure of DNA. At one point, after lunch with Kreisel, they popped into Blackwell's bookshop to find a copy of Pauling's textbook to check some facts about bonds.

Back in Cambridge on Monday morning, they began to fiddle with models of metal atoms and wire bonds borrowed from Kendrew. They had to improvise the large phosphorus atoms by wrapping wire around carbon atoms. After gooseberry pie at the Eagle, they began building a model in earnest. By the evening, and before dinner at the Green Door, they had assembled a three-chain structure with the phosphate-sugar backbones on the inside and the bases sticking out. Why three chains? Because the density of the crystals implied at least two and more likely three chains per molecule. Why with the phosphates on the inside? Because Watson had remembered Franklin saying that there were only eight molecules of water in each unit cell, and the negative electrical charge on the phosphates (implied by the acidity of DNA) must therefore need close association with positively charged metal ions. Where better to put the ions than on the inside? Watson made the wild assumption that hitherto unsuspected magnesium ions lay within the core, holding the chains together. This gave a helical structure with roughly the right dimensions. On Tuesday, Crick telephoned Wilkins, who travelled up to see the model on Wednesday. He came with four others, including Franklin: panic had set in at King's at the news that there was a model at Cambridge.

When they arrived, and after a short lecture from Crick on helical diffraction theory, Franklin took one look at the model and declared it worthless. Far from saying eight molecules of water per unit cell, she had said eight per lattice point: 24 times as many. A good physical chemist, Franklin knew that every metal ion must be surrounded by water molecules; it could not be naked like the magnesium ions in the structure before her. She was sure that the phosphates were on the outside because the switch, when the fibre was wetted, from crystalline (A) to paracrystalline (B) implied that water freed the sodium ions from between—rather than within—the DNA molecules. The group from King's did little to conceal their contempt. Even Crick's ebullience faded after they all went for lunch at the Eagle: Gosling recalls to this day the "delicious moment" when Crick and Watson were reduced, for the first time in living memory, to silence.

They had trespassed on others' turf and been utterly humiliated. Wilkins soon wrote to Crick, politely asking him to leave DNA alone; and it was inevitable that Bragg would speak to Randall, agree to call Crick and Watson off DNA altogether, and order Crick to get back to his thesis. Even the parts used to make the models were handed over to King's as a sort of apologetic gesture. Not that Franklin, or even Wilkins, had any time for such a childish approach as building models. The only eager model builder at King's, Bruce Fraser, had just left for Australia. Besides, the last two attempts at model-building at the Cavendish—first by Bragg, Perutz and Kendrew; then by Crick and Watson—had been fiascos. Franklin firmly believed that trial and error was an outdated approach in crystallography, and that pure induction was the way to go.

In early 1952 Maurice Wilkins tried to lift his own melancholy mood by travelling to see his German girlfriend in Munich, then going on to Bern in search of more "Signer DNA," and then returning to Naples for some more cuttlefish sperm. On the train from Innsbruck to Zurich, he wrote a letter to Crick:

> Franklin barks often, but doesn't succeed in biting me. I won't start making any references to the "business" between you people and us over n.a. but look forward to discussing all our latest ideas and results with you again. . . . I have found several of your suggestions very valuable, but am fairly convinced for many reasons the

phosphates must be on the outside. I hope Bragg neither barks nor bites.

He drew in the margin of the letter a diagram showing the pattern he had gotten from better X rays of cuttlefish sperm. It showed an X-shape pattern of layer lines, much like the best of Rosalind Franklin's photographs of the B form. It screamed "helix." But Crick was no longer doing DNA.

Chapter Five

Triumph

As 1952 began, Rosalind Franklin, together with the student she had inherited, Raymond Gosling, now had an effective monopoly on the search for the structure of DNA. Wilkins, Watson, and Crick had all been seen off in different ways. Other X-ray labs were not in the race. Astbury and Hodgkin were doing other things. In Bernal's lab at Birkbeck, the only person working on DNA, Sven Furberg, had gone back to Norway after brilliantly working out that the plane of the bases was perpendicular to the plane of the sugars and then trying to build a single-helix model. Only Linus Pauling in distant California was still free to speculate, but he had no access to the X-ray data—Randall had refused Pauling's rather presumptuous request to send copies of Wilkins's pictures. When Pauling tried to come to London for a meeting about proteins in May 1952—at which point he would surely have wanted to visit King's—his reputation as an outspoken antinuclear pacifist caused the U.S. State Department, under pressure from Senator Joe McCarthy, to revoke his passport.

True, there were still biochemists working on the chemistry

of DNA. Just across Cambridge the biochemist Alexander Todd had by now worked out the precise links in the chain of sugars and phosphates. Each phosphate was attached to the third carbon atom in one sugar molecule and to the fifth in the next. This 3-5-3-5 . . . pattern gave the backbone of a DNA direction—either the 3 was up and the 5 down or vice versa. The most prominent biochemist working on DNA was Erwin Chargaff at Columbia University, a rather grand Austrian émigré, who had found an intriguing fact about its nitrogenous bases. Although the proportion of each base varied from species to species, there was usually a neat symmetry: the amount of adenine was the same as the amount of thymine, and the amount of cytosine was the same as the amount of guanine. Chargaff had no idea what his ratios meant.

Chargaff came to Cambridge in the last week of May 1952, and Kendrew invited Watson and Crick to meet him after lunch at Peterhouse College. The meeting was a disaster. To the dignified, erudite Chargaff, the "undeveloped" Watson was bad enough, but the ebullient Crick was positively offensive: "The looks of a fading racing tout; something out of Hogarth (*The Rake's Progress*); Cruikshank, Daumier; an incessant falsetto, with occasional nuggets glittering in the turbid stream of prattle." It became quite clear that Crick did not know thymine from cytosine. Nor had he even heard of Chargaff's base ratios. Chargaff was left with the impression of "a typical British intellectual atmosphere, little work and lots of talk." Later, bitterly realising how close he himself had come to discovering base pairing, he became an entrenched critic of everything about molecular biology: "That in our day such pygmies throw such giant shadows only shows how late in the day it has become."

But for Crick, the news of Chargaff's base ratios was a bolt from the blue, because, trespassing back into DNA, he had been thinking about how the bases might pair up if they were on the inside of the helix. He had imagined them layered on top of each other, somehow spelling out a message, and that message somehow getting copied. He was envisaging like-with-like pairing so that the sequence of one set of bases would be copied directly. He had discussed it in the pub with a young mathematician training to be a biochemist, John Griffith, who played with the bases and came back with the news that adenine should attract thymine and guanine should attract cytosine. Fine, thought Crick; the replication could be complementary. Copying one message gave another, which, copied, gave back the first: A makes B, and B makes A—like a photographic negative, or a lock and a key.

So when Crick heard of Chargaff's ratios (adenine = thymine; guanine = cytosine) he rushed off to check that they were indeed the same pairings as Griffith had suggested. For a week at the end of July, while Watson was away at a meeting in Paris, Crick even tried a laboratory experiment to see if he could detect base pairings in solution. The rationale was that if the bases were paired, they would absorb less ultraviolet light. But the experiment failed (the effect was much too weak to be detectable), and he went back to his proteins. As it happens, Griffith was right for entirely the wrong reason. He and Crick were thinking of pairing the bases interleaved, flat side to flat side, whereas in fact they pair end to end, like dominos. Nonetheless, the idea of a sequence of bases being copied through a negative version of the sequence had entered Crick's mind.

That summer Crick met Rosalind Franklin for the second

time when she came to a meeting in the zoology department. She told him, as they waited in line for tea, that she now thought the A form of DNA was not helical at all. The X rays in one photograph were too asymmetrical: the zigs were stronger than the zags. She seems to have come to believe that the drier A form was an unwound version of the B form, just as Pauling had found that alpha-helix proteins could be unwound into so-called beta sheets. This had led her to announce on 18 July the "death of DNA helix (crystalline)" in a mock funeral notice addressed somewhat pointedly to Wilkins, who was reluctantly convinced by her argument. In the line waiting for tea, in August, Crick was not. Unready as always to let one piece of data spoil a good theory, he argued that the asymmetry was misleading: it could be caused by slight differences in the parallel packing of different DNA molecules in each crystal—and have nothing to do with the structure of each molecule. His intuition, it later emerged, was right. But Franklin was now set on the course of slowly and painstakingly extracting information about the A form by cylindrical Patterson superposition—a task that would take months of calculations—and as usual she believed in letting the facts speak for themselves rather than imposing guesses on them.

In the autumn of 1952, the Cricks moved to a new house, 19 Portugal Place. Uncle Arthur, the pharmacist who had paid for Francis's graduate education at UCL, gave them the money to buy it. It was one of a pair of tall, narrow attached houses in a quiet pedestrian lane just around the corner from the twelfth-century round church on Bridge Street and next to a modern pub called the Maypole. Five steep stone steps led up to the front door, and three timbered box-sash bay windows jutted out one above the other. With an attic and a basement, the house had five floors, but it was very narrow. Since the birth of Gabrielle in July 1951, Michael had been staying with his grandmother, Annie Crick having sold her house in Northampton and moved to a large house on Barton Road in Newnham on the western outskirts of Cambridge. She lived in the ground floor, renting out the other two floors as flats. Annie and her sister Ethel had chosen the boarding school Dunhurst (the junior school of Bedales) for Michael, sharing the fees between them.

Family life could now be more social. Watson came around for meals, especially on Sundays, and for advice on how to kick-start his nonexistent love life. Odile, who had been giving lectures on the history of costume at the Tech (now Anglia Polytechnic University) before the baby came, had effortlessly amassed a penumbra of bohemian artistic friends; and Francis brought intellectuals from Caius College, where he had dining rights. Sometimes given to wearing colourful waistcoats, Crick was a dandyish feature on the Cambridge scene who clearly took trouble over his appearance. This was unusual in the scientific world; not many academics had a subscription to *Vogue*. Indeed, Crick stands out among the great scientists of history precisely because he was not eccentric, silent, shy, or obsessive. He was gregarious and an extrovert.

As Watson put it, once Crick had beaten Pauling to the coiled coil that autumn, "There was growing acceptance both in and outside Cambridge that Francis's brain was a genuine asset. Though a few dissidents still thought he was a laughing talking-machine, he nonetheless saw problems through to the finish line." So it was that he suddenly received an invitation to go, sometime next year, to the Polytechnic Institute of Brooklyn, in New York, for a year—and a welcome $6,000. There David Harker was gathering a team to X-ray proteins, and he had heard that Crick was one of the best. Crick accepted the offer and applied for a visa to the United States.

Meanwhile, Watson and Crick had two new American colleagues to share their large room at the Cavendish. One was Jerry Donohue, a former student of Linus Pauling. The other was Peter Pauling, son of Linus, but unlike his serious father more intent on cutting a swathe through Cambridge's girls than through its ideas. In the week before Christmas, Peter Pauling startled Crick by announcing that his father had written to say that he had solved the structure of DNA by building a model and was sending a paper to be published. This was a nightmare come true. The year of the moratorium, when King's had used its effective monopoly on DNA to little effect, had allowed Pauling to catch up. Perhaps he had even beaten them again. Peter wrote to his father that for more than a year Crick had been invoking the ogre of Pauling to scare King's into action. Now the ogre had come.

It was not until 28 January 1953 that Pauling's manuscript reached the Cavendish. Both Peter Pauling and Bragg received copies. As Watson and Crick skimmed through Peter's copy, despair gave way to hope again. The structure Pauling had proposed was eerily similar to the one they had built more than a year before: it had three chains wound around one another, with the bases sticking out horizontally. However, far from solving the problem of tight packing with no water but imaginary magnesium ions in the middle, Pauling had made an even worse blunder. He had packed the core even tighter, and with hydrogen bonds, between non-ionised phosphates. This made no chemical sense—it made DNA not an acid.

Two days later Watson enacted a now legendary scene at King's: he showed Rosalind Franklin Pauling's manuscript, irritating her to the point where he feared she might hit him; retreated with Wilkins to another room; and stood amazed as Wilkins showed him a near-perfect B-form photograph that Franklin and Gosling had taken the previous May. Wilkins had the photograph because Franklin was by now preparing to leave King's for Birkbeck College and had handed photographs, project, Gosling, and all back to Wilkins.

Watson rushed back to Cambridge with the news of an unambiguously helical form of DNA—something Crick should have known from the diagram Wilkins had sent him nearly a year before in the letter written on the Austrian train. But Watson came back with another fact, possibly gathered from Wilkins over dinner on the evening of the scene, and surely not just glimpsed from the photograph. At the top as well as the bottom of the photograph was a very dense black smudge, which lay precisely on layer line 10. This meant that each helix must have 10 nucleotides per turn—10 phosphates, 10 sugars, and 10 bases. Since the distance between two nucleotides was 3.4 angstroms, the pitch of the B-form helix was not 27 angstroms but 34.

Until now it was Watson who had kept going to King's and had remained in touch with Maurice throughout 1952. The one occasion during this period when Crick went to have lunch with Wilkins, intending to discuss base pairs, he failed to bring up the subject at all. It was Watson who now went to Bragg and asked for the machine shop to make models for them so that they could start playing again. Bragg knew what this meant: the moratorium was over. Since King's was incapable of fending off the ogre—Pauling—Watson and Crick must be unleashed. And it was Watson who during the first week of February began to assemble models of two-chain DNA. Over Crick's doubts, he insisted that it was two chains. He had seen, as Crick had not yet, that the shrinkage of the B form to the A form revealed quite clearly that there must be two chains, not three. Besides, Watson said, biological things come in pairs. At first Watson persisted with internal phosphate backbones. When he complained that they were not working, Crick asked him why he did not put the phosphates on the outside, to which Watson replied that this would be too easy. All the more reason to try it, said Crick, pretending to concentrate on his thesis.

On Sunday, 8 February, Wilkins came to lunch at Portugal Place. Peter Pauling and Jim Watson were also there. They spent the meal trying to persuade Wilkins to begin urgently building models. He said he would not do so until Franklin had left in March. Then, at some point in the afternoon, Crick posed the question directly: "Then do you mind if we have a go?" In fact, they had already started. There was a long pause as poor, tortured Wilkins saw his chance at regaining control of the DNA story slip away. He found the question "horrible." But he consented.

A few days later Crick obtained from Max Perutz a short report written for the MRC the previous December on the work at King's. Much ink would be spilled more than a decade later by Perutz, Bragg, and Randall about whether this was a breach of trust on Perutz's part. Perutz maintained that the report was not marked "confidential," and indeed was intended for circulation within the MRC; Randall maintained that it should nonetheless have been treated as private. It was vital to Crick, but only because it told him things that had already been aired at public meetings where he happened not to have been—the colloquium in 1951 in particular. It contained the following sentence, written by Franklin: "It was apparent that the crystalline form [A] was based on a face-centred monoclinic unit cell with the C-axis parallel to the fibre axis." There followed her estimate of the dimensions of the unit cell. This told Crick at once, as the seminar of November 1951 would have told him, that the two chains ran in opposite directions, because to be monoclinic the structure had to look the same upside down as right side up. "This was the crucial fact," Crick later told Horace Freeland Judson. "Furthermore, the dimensions of the unit cell, which were also in the Council report, proved that the dyad had to be perpendicular to the length of the molecule and implied also that the duplication was in fact within the single molecule."

Now came Crick the visualiser. With the chains running in the same direction, the structure repeated itself after half a turn of each helix. If the chains were going in opposite directions (i.e., one went from sugar atom 3 to sugar atom 5, and the other went from 5 to 3), then the repeat came after a whole turn of each helix. So the 10 nucleotides could be fitted into an ample 360 degrees, not a tight 180, and the angle between each sugar and the next could be 36 degrees, not 18. Watson either could not or would not get the point, so while Watson was playing tennis one afternoon, Crick rebuilt the model himself and left a note on it: "This is it—36 degrees rotation." This was as close to a "eureka moment" as they had had, and it was all Crick's.

There was one more eureka moment to come, and it would be all Watson's. Crick and Watson still had no idea how to fit the bases into the middle of the structure. Watson had begun to realise that the bases could form hydrogen bonds with each other end to end, domino fashion. Crick missed this, mainly because he thought that the atomic configuration of the bases could switch at random between different tautomeric forms, and he could not see how hydrogen bonds could exist between unstable structures. He was making an elementary chemical error: tautomers are alternative arrangements of atoms, but each one is perfectly stable. Watson seems not to have gotten hung up on this problem. Watson now became excited at the idea of like-for-like pairing, where adenine on one chain paired with adenine on the other. Jerry Donohue then looked up from his desk and said that Watson was using the outdated and less likely—indeed wrong—tautomer for each base, the "enol" instead of the "keto" configuration. Watson, unwilling to wait for new keto base models from the machine shop, then designed and cut out cardboard ones instead. He finished on Friday evening, 27 February, and went home.

Saturday, 28 February, was a fine spring day. The crocuses were flowering along the banks of the River Cam. Watson came in to work before the others and began playing with his cardboard bases. Quite suddenly, he saw something that once seen would never be unseen again. Adenine paired with thymine, separated by the distance of parallel hydrogen bonds, was exactly the same shape as cytosine paired with guanine. Each base pair, being the same shape as the other, could go anywhere in the core of the helices.

Donohue had come in while Watson was still working this

out. When Crick sauntered in at mid-morning, Watson nervously showed him the base pairs. Crick quickly saw that they must be right, for two reasons. First, they explained Chargaff's ratios; and second, they had the right symmetry: the bond attaching each base to its chain was 90 degrees from the bond attaching its partner to the other chain. This symmetry meant that each base in a pair could be flipped to the opposite chain, but only by turning it upside down at the same time, thus proving that the chains ran in opposite directions. Again, Crick's famous visualising power was at work. Everything was now in place, and far from proving sterile like the alpha helix, the structure revealed profound insights into the nature of life itself. Here was a code of infinite possibility. The bases could go on either strand in any order, but any message written on one strand must produce a complementary copy of itself on the other strand simply by the base-pairing rules. The two helices could unzip down the hydrogen bonds and duplicate their codes. Heredity was manifest in the very structure.

They went to the Eagle for lunch and, according to Watson, over a pint of bitter Crick announced out loud to everyone in hearing that they had found the secret of life. Crick did not later remember this, but he did remember telling Odile that night that they had made a big discovery. She took no notice: "He was always saying things like that." But whereas Crick was confident, Watson felt queasy. Over the following weeks it was Watson who lived in terror that they would be wrong, and Crick who knew without a shadow of doubt that they were right. "The funny thing was how nervous Jim was about the structure. He didn't like me explaining it to people."

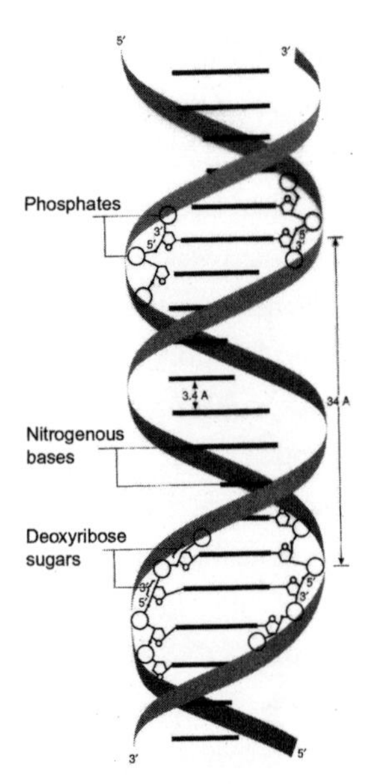

The Double Helix

Explaining it to people was what Crick now did all day. Perutz, Kendrew, Bragg, the biochemist Todd, various physicists, and others were summoned to see it. As soon as the machine shop had produced the model bases, made of flat plates of galvanised metal with narrow brass tubes for bonds, Crick painstakingly rebuilt the model with plumb line and ruler ("Jim wasn't much good at that sort of thing"), finishing, exhausted, on Saturday, 7 March—whereupon he went straight home to bed. The model stood, several feet high, on a table in their room, each flat base clumsily held in place by clamps attached to a vertical support. That very day, unknowing, Wilkins wrote a letter to Crick: "Our dark lady leaves us next week. . . . At last the decks are clear and we can put all hands to the pump. It won't be long now." It was Kendrew who called Wilkins and told him. Wilkins came on 12 March, and the mood turned sour. He knew at once that the structure was too perfect to be wrong, but he made no secret of his bitter disappointment that his two friends had finished the job the very weekend before he was going to start building a model himself. He refused their offer of joint authorship on what would soon become the most famous paper in all of biology.

It was eventually agreed by Bragg and Randall with the editors of *Nature* that three papers would be published simultaneously: one by Watson and Crick (the order of the names was decided by a coin toss); one by Wilkins, Stokes, and Herbert Wilson; and one by Franklin and Gosling. Watson and Crick's paper was drafted by Crick; typed by Watson's sister Betty, who was living in Cambridge at the time, and illustrated with a simple pair of intertwined ribbons linked by slim bars, drawn by Odile. Back in London, on 18 March, having received a draft of their paper, Wilkins calmed down enough to write, "I think you're a couple of old rogues but you may well have something." He added in a later card, "Could you delete the sentence 'It is known that there is much unpublished exp[erimental] mat[erial].' (This reads a bit ironical.)" The paper was sent to *Nature* on 2 April.

It is not entirely clear when Franklin first saw the model. She drafted a paper about her own work on 17 March, after moving to Birkbeck on Saturday 14 March. On 10 April she wrote to Crick asking if she could bring Gosling to see the model the following Tuesday, 14 April, and this was probably her first sight of it. She immediately saw that the laboriously calculated Patterson analysis confirmed the model in every particular. When

Crick saw Franklin and Gosling's manuscript, he was astonished by how well the data in it confirmed the model. It proved that the phosphates and sugars were on the outside and that there were two chains; and it included the famous photograph that Watson had glimpsed but Crick had never seen, which dramatically confirmed the C2 space group by the absence of any spot (through destructive interference) on the fourth layer line. This told Crick—who applied his visual intuition again—that vertically, the two chains were not spaced apart equally. The gaps between them were three-eighths and five-eighths of a period. "With her data we built a better model."

At this, even Watson's queasiness began to settle. He had wanted no speculative mention at all of the implications for heredity, eventually agreeing only to the famously enigmatic sentence "It has not escaped our notice that the specific pairing we have postulated immediately suggests a possible copying mechanism for the genetic material." Now, with Franklin's proof in hand, a second and bolder paper was drafted about the genetic implications of the structure of DNA. It is generally assumed that Crick wrote this second paper, but the manuscript that survives is in Watson's hand, with the figure legends and a few sentences added by Crick. This may explain why it still failed to mention the argument about C2 symmetry, one which Crick relied on but Watson did not fully understand.

Watson gave a talk about the double helix at the Hardy Club (an informal group of like-minded biologists and physicists at Cambridge) on 1 May; too much good Peterhouse wine reduced him at the end to murmuring, "It's so beautiful, you see, so beautiful." On 21 May, an undergraduate aspiring to be a journalist sent a freelance photographer, Anthony Barrington Brown, to take pictures of the two young men and their metal model for a possible article for *Time* magazine. Barrington Brown, himself a former chemistry student, found the two in a jovial mood, and it was all he could do to pose them for a formal shot. He asked them to stand by the model and look portentous, which they "lamentably failed to do, treating my efforts as a bit of a joke." He eventually persuaded Crick to stand on a stool, using a slide rule to point to a feature of the model, while Watson, clothed by Odile in a new jacket for the occasion, looked on from the other side. *Time* magazine never used the photographs, but it paid Barrington Brown half a guinea for them. One of them is now among the most famous photographs in all of science.

It was a momentous spring: Everest climbed, Elizabeth crowned, Stalin dead, *Playboy* born. The biggest event of all—life solved—caused barely a ripple. Bragg announced the news at a Solvay conference on proteins in Belgium that began on 8 April, but it went unreported. Bragg again mentioned it on 14 May at a meeting in London, and this time it was picked up by Ritchie Calder in the *News Chronicle* the next day under the headline "Why You Are You: Nearer Secret of Life." Calder's article concluded: "Discovering how these chemical cards are shuffled and paired will keep the scientists busy for the next fifty years." The news may have reached readers of the early edition of the *New York Times* the following day, under the baffling headline "Form of 'Life Unit' in Cell Is Scanned," but this item was apparently pulled from later editions. In June one Sunday newspaper also carried a short piece, in which Crick was briefly quoted. Otherwise, there was no coverage.

At the end of May Watson left for America, with a new, portable model of the double helix made by Tony Broad, the engineer, as well as a paper he was to read at Cold Spring Harbor in June—the first seminar that either he or Crick had been asked to give about the secret of life. A personal invitation from Linus Pauling asking Crick to come to a conference in California in September, "to speak as much as possible during the meeting," turned out to be a hoax devised by Watson and Peter Pauling. Crick took it well; Odile did not; Linus Pauling docked Peter's allowance by £5. Pauling's reaction to the double helix was initially cautious, even slightly defensive. He had written to Crick in March, saying he thought it "fine that there are now two proposed structures for nucleic acid, and I am looking forward to finding out what the decision will be as to which is correct." But after seeing the model in April he soon realised what the answer was.

"If Watson had been killed by a tennis ball," Crick wrote 21 years later, "I am reasonably sure I would not have solved the structure alone, but who would?" This is a question that still hangs in the air. At the time Crick thought Pauling would have found the structure, but Pauling might not have reconsidered his own model in time. Wilkins, with Gosling's help, was about to build a model that March. He probably needed—and would have sought—Crick's help to see the antiparallel chains, but he would surely have seen the base pairing on his own. He, not Franklin, is the one most cheated of destiny by Watson's haste. Sir Aaron Klug's exhaustive survey of Rosalind Franklin's notes leaves little doubt that she moved towards seeing both the antiparallel chains and the base pairing in early 1953, but did not reach either insight. She then left for Birkbeck on 14 March, and Randall firmly told her to leave DNA behind at King's. She did not entirely do so, because she continued to supervise the completion of Gosling's thesis, and she wrote up a paper on the A form, which proved to be a compressed version of the double helix with the bases at angles. Nonetheless, the manuscript she wrote on 17 March, still two crucial steps short of the answer, might well have been her last word on the subject. Her tragedy is not that she nearly got there but that she could have gotten there the year before. As Gunther Stent argued, the answer to Crick's question about the tennis ball is probably that the solution would have emerged gradually from several people.

The story of the double helix is awash with might-have-beens. Every participant had cause for regret about a blunder made or an opportunity missed. Randall sowed fatal confusion between Wilkins and Franklin, which ensured that they never collaborated as Watson and Crick did. Wilkins should have built models sooner. Franklin should have learnt more crystallographic analysis or shared her thoughts with somebody. Watson should have taken notes. Pauling should have heeded elementary chemistry (or been treated less unreasonably by the State Department). And Crick should have tried harder to befriend Franklin; they later became good friends. Even the minor players in the drama could kick themselves. Sven Furberg and Bruce Fraser gave up prematurely on building models. Astbury, Bernal, and Chargaff simply never imagined that the structure would prove so revealing, so they never tried to model it. Yet each made an invaluable contribution, and Crick and Watson were in one sense lucky to place the keystone in the arch—or the last clue in somebody else's crossword. Still, as Crick wrote, "It's true that by blundering about we stumbled on gold, but the fact remains that we were looking for gold."

To the end of his life, Crick would insist that what mattered was the discovery of the double helix, not who made it:

> Rather than believe that Watson and Crick made the DNA structure, I would rather stress that the structure made Watson and Crick. After all, I was almost totally unknown at the time and Watson was regarded, in most circles, as too bright to be really sound. But what I think is overlooked in such arguments is the intrinsic beauty of the DNA double helix. It is the molecule which has style, quite as much as the scientists.

Crick once told a newspaper reporter (in Hawaii): "Unlike the jet engine, which had to be invented, the DNA structure was always there." Scientific discoverers are dispensable in a way that artists are not. Gravity, America, and natural selection would all have been discovered by somebody else if Newton, Columbus, and Darwin had not gotten there first, whereas nobody would have written *Hamlet*, painted the *Mona Lisa* or composed the Ninth Symphony if Shakespeare, Leonardo, and Beethoven had not done so. Yet it is precisely because scientists have to be first that their achievement is even more remarkable. Shakespeare did not have to beat Marlowe to the first draft of *Hamlet*.

Chapter Six

Codes

"MY DEAR MICHAEL," wrote Crick on 17 March 1953 to his 12-year-old son at school. "Jim Watson and I have probably made a most important discovery." He went on:

> Now we believe that the DNA is a code. That is, the order of the bases (the letters) makes one gene different from another gene (just as one page of print is different from another). You can now see how Nature makes copies of the genes. Because if the two chains unwind into two separate chains, and if each chain then makes another chain come together on it, then because A always goes with T, and G with C, we shall get two copies where we had one before. In other words, we think we have found the basic copying mechanism by which life comes from life. . . . You can understand that we are excited.

This is the earliest written description of the genetic mechanism, written before the second paper for *Nature* was drafted, and it is correct in every particular. Matters such as how the two chains could unwind, and whether the base pairing was spontaneous or needed help from some protein machinery, would detain scientists for many years yet. Indeed, proof of semiconservative replication did not appear for five years, and the precise structure of the double helix was not actually proved beyond doubt until the late 1970s. But Crick's letter to his son sets out the truth that now lies at the heart of biology, and that was utterly unsuspected until 1953: the existence of a digital cipher that can be automatically copied. Like all discoveries, though, it posed a much bigger question than it answered. How is the code used? What is it a code for? The next 13 years of Crick's life would be dominated by these questions, and answering them would be his triumph. For although the double helix made Crick, to a large degree Crick made the genetic code. He set the terms and shaped the debate; he also guessed much of the answer. Though the result was less of a surprise, it was in many ways a greater scientific achievement than the double helix.

There was little disagreement about what the code must do: it must translate a sequence of bases of DNA into a sequence of amino acids of protein. This was just a guess, but it was obvious, and right. Proteins do all the work in the body, and like DNA they consist of long, unbranching chains of similar units. George Beadle's famous experiments with bread mould in 1941 showed that one genetic mutation affected one particular protein. As Crick wrote, "The main function of the genetic material is to control (not necessarily directly) the synthesis of proteins. . . .

Once the central and unique role of proteins is admitted, there seems little point in genes doing anything else." One day during that summer of 1953, Watson and Crick sat down in the Eagle and wrote out the canonical list of amino acids known in proteins, carefully discarding many slight or rare variants in obscure proteins that biochemists had been collecting like stamps over the previous few years. They came up with 20. That they got the list exactly right, despite being amateur biochemists, is a minor miracle.

There was a gentle irony in the fact that Michael Crick was the first to read about the genetic code. In 1950, when Michael was 10, his father had given him a book called *Codes and Ciphers* and had challenged him to write a code that Francis Crick and Georg Kreisel could not crack. Michael promptly devised a code that defeated not only the country's future greatest biologist but also one of its leading mathematical philosophers. This code contained degeneracy: in other words, there were several different ways of encoding the same letter. Later, the genetic code would also prove to have degeneracy. (Michael Crick would go on to be a pioneer in computer software, as would Michael's son Francis and daughter Camberley.)

Crick did not immediately get to work on the genetic code. First, he had to finish his thesis on the structure of haemoglobin. He toyed with the idea of replacing this with a thesis on DNA, but he could not disentangle his own contribution from that of Watson, though he added the two papers on DNA as an appendix. At last he got his doctorate, in July, not for the secret of life but for this rather thin conclusion:

> The work on haemoglobin presented here, both theoretical and experimental, represents rather the clearing of the ground for a further attack. . . . Nevertheless, it leads to a concrete hypothesis: that globular proteins largely consist of lengths of helices—probably alpha—packed together in a non-parallel manner.

In other words, the structure of protein was complicated and irregular.

Then on 22 August, with Odile, Michael, and Gabrielle, Crick embarked on the *SS Mauretania* at Southampton and sailed for New York to take up his year's fellowship at Brooklyn Polytechnic for further work on protein. It was to prove a mostly depressing and lonely experience. The X-ray work on ribonuclease was mundane and unrewarding. Apart from Bea Magdoff, a good crystallographer who helped him develop the theory of isomorphous replacement; and Vittorio Luzzati (an old friend of Rosalind Franklin's), Crick found few like-minded colleagues at the Brooklyn Polytechnic. He got on well enough with Dave Harker, the head of the lab; and with Harker's Russian wife, Katherine, the daughter of a czarist prosecutor, but they were not scientific gossipers. Odile was especially disappointed. She had been brought up to think of America as Hollywood, and she found life in apartment 610 at 9524 Fort Hamilton Parkway in Brooklyn "pretty terrible." Frilly lampshades in a condominium in an outer borough of New York were not her style. She was also pregnant. Only Michael, in a Brooklyn high school for a year, enjoyed himself.

The worst of it was that the Cricks were short of money.

More than a quarter of Crick's monthly salary of $450 went towards rent, and other expenses were also high. This led to a fairly serious rupture in his relationship with Watson, one that would contribute to Crick's strong reaction to Watson's book *The Double Helix* 12 years later, when the shoe would be on the other foot. Watson was now at the California Institute of Technology in smoggy Pasadena, where, despite the presence of Pauling, Delbrück, and Richard Feynman, he too was dissatisfied with life. Crick and Watson were both missing their conversations together. Crick had been approached to do a series of broadcasts for the BBC Third Programme (the highbrow radio channel), which had picked up on the fact that something important had happened. But Watson thought that Crick should not do this, as it would come across as boastful to talk about DNA on the radio; and Bragg said that Crick should not do it without Watson's agreement. From Brooklyn Crick wrote to Watson: "Do you still feel you can't allow the Third Programme Broadcast? I've yet to find anyone who will say they would object to it, and things have cooled down a bit now. Also, it would bring in $50 to $100, which at the moment I could do with."

Watson's reply from Pasadena was harsh:

> Concerning the BBC. I still think a talk on the 3rd would be in bad taste. There are still those who think we pirated data. . . . Judging it on a monetary basis ($100) seems unfortunate. Basically, however, you are the one to suffer most from your attempts at self publicity. My main concern is not to be dragged into it as I'm afraid I was in Cambridge. If you need the money that bad, go

> ahead. Needless to say, I should not think any higher of you and shall have good reason to avoid any further collaboration with you.

Crick replied a few weeks later:

> As you were so set against it I did not allow the BBC to broadcast in the Third, although your name is mud in the Crick household because of this. However, I did write an article for Discovery, as the Editor's argument (crudely, that if I didn't do it, somebody else might, and that would be worse) seemed to me to be more to the point than yours. [George] Gamow hinted that the Scientific American would like an article. How do you feel about this? You should realise that as a married man with two children (+) I cannot afford to take your detached attitude about money. We live very quietly here mainly because we are so broke.

To rub salt on the wound, Watson posed for *Vogue* that summer, appearing in a photographic essay on talented young Americans. In retaliation Crick did write for *Scientific American* and did eventually do two BBC broadcasts in November and December 1955. It is impossible to see, now, what Watson could have objected to. Like much of the output of the BBC at the time, these broadcasts were learned lectures, slightly patronising in tone—the double helix is a spiral staircase with split steps—and very cautious about speculating beyond the known facts.

The argument about the BBC did not prevent Watson and

Crick from writing each other enthusiastic letters, dense with both gossip and scientific argument, for some years to come, until these communications petered out in planned itineraries and flippant postcards. (One memorable postcard from Crick to Watson in 1957 read simply: "Are you dead? Or in love? Francis.") In these early years their correspondence retained an edge of competition, but also complicity: they felt alone in a world of uncomprehending idiots. The reactions of many biochemists to their structure for DNA "ranged from coolness to muted hostility," Crick would later write; and geneticists barely noticed it at all. To have announced the secret of life to the world and receive so little reaction was baffling.

In February 1954, rather than remain in Brooklyn, Odile and Gabrielle travelled home to stay with Odile's mother in King's Lynn for the birth of the new baby. Jacqueline was born on 12 March 1954, her mother resisting Crick's suggestion (seconded by Watson in California) that she name the child Adenine. Francis and Michael moved to an apartment in Brooklyn Heights—"not cheap, but better for my morale." Crick stayed on till July, giving a lecture series on DNA at the Rockefeller Institute and visiting Chargaff, who predictably said that Crick's speculations about genetics were all wrong. Crick then moved to Woods Hole for the month of August to meet Watson and the physicist George Gamow for their first attack on the coding problem.

George Gamow was a hard-drinking Russian émigré physicist famous for his new theory of the big bang; and for his lighthearted popular books about science, including the newly published *Mr. Tompkins Learns the Facts of Life*. He had written to Watson and Crick out of the blue in July 1953 after reading the second paper on DNA, immediately seeing the significance of their discovery: "Each organism will be characterised by a long number written in a quadrucal [sic] system with figures 1, 2, 3, 4 standing for four different bases." Gamow was now at the centre of speculations about coding and dragged several eminent physicists on the west coast into his quest, devising his first code with Melvin Calvin and Edwin McMillan in Berkeley, then another in Pasadena with Richard Feynman—until, in Gamow's inimitable style, "Dick Finemann succeded to show that no solution excist." (It later emerged that Gamow's spelling was just as erratic in his native language, Russian.) Gamow then made his third attempt with Edward Teller, the father of the hydrogen bomb. Biochemists and biologists might not notice the double helix, but Gamow made sure that the physicists did.

For all Gamow's enthusiasm, his ideas came to nothing at Woods Hole that August, though Watson pulled off a good practical joke by inviting everybody to a "Wiskie Twistie RNA party" in Gamow's name at Gamow's cottage without telling him. Crick torpedoed all Gamow's codes with one neat fact. (This time he was ready to let one piece of data spoil a good theory.) The sequence of amino acids in the protein insulin was gradually being deduced by Fred Sanger in the biochemistry department at Cambridge, an extraordinary achievement for which this self-effacing man would win the first of his two Nobel Prizes, and Crick had befriended Sanger and seen most of the sequence. It already showed that any amino acid could apparently have any neighbour. Gamow's ideas were all based on the shapes of the "holes" in the grooves of the double helix, and because they suggested triplets of bases overlapping by two, they all demanded some kind of constraint on what could go next to what. Gamow's scheme allowed only eight amino acids to have as many as seven different neighbours each. In insulin alone 10 amino acids have eight neighbours or more.

At Woods Hole, Watson persuaded Crick and Gamow that they must switch their attention to RNA, the other kind of nucleic acid with one extra oxygen atom in each of its sugars. If DNA could not directly specify protein structure, perhaps it used RNA as an intermediate. Unlike DNA, RNA did not live exclusively in the cell nucleus but seemed to be everywhere in the cell, in a bewildering variety of sizes. Watson, in a letter to Delbrück in 1952, had predicted that the job of RNA was to mediate between DNA and protein; and in 1953 he had begun a search for its structure though without success so far. While driving along a Californian freeway one day in the spring, Watson and the chemist Leslie Orgel had conceived the idea of a club for people interested in RNA, an idea Gamow had quickly taken over. The "RNA Tie Club" was to have 20 members, each one with a necktie adorned with a squiggly RNA and a unique tie pin bearing the abbreviation for one of the 20 amino acids. The club had a motto—"Do or die or don't try"—and several officers, including Synthesiser (Gamow), Optimist (Watson) and Pessimist (Crick).

On 8 September 1954, Crick sailed from New York. Reunited with Odile and meeting Jacqueline for the first time, he resumed family life at Portugal Place. In King's Lynn with her French grandmother, little Gabrielle had become fluent in French. Crick had a job in Cambridge again, though it was only a seven-year contract from the Medical Research Council. This was possible because Bragg had left to head the Royal Institution, and Perutz wanted Crick back for his expertise in protein structure. Crick had been briefly tempted by a permanent position at Edinburgh University, where his friend the zoologist Murdoch Mitchison had just gone; but Cambridge was where he wanted to be, whatever the job. David Blow, a student of Perutz, remembers the sudden arrival of the extravert Crick: "At first sight he was just going to be a nuisance, as his method of working was to talk loudly all the time." But Blow then asked Crick's advice and was astonished at the speed with which Crick grasped, then solved, a problem that had baffled Blow for weeks: how to minimise the errors in the electron density map of an isomorphous replacement experiment in haemoglobin. The joint paper that eventually resulted would transform protein crystallography.

Blow also noticed, though, that proteins could no longer hold Crick's attention. What Crick mainly talked about now was genes, and the coding problem in particular. Just before he left America, he had attended a Gordon Conference in New Hampshire at which he gave a talk about DNA. Driving south, he had an idea. Later, back in Cambridge, he wrote it down in a rather gloomy paper he prepared for circulation to the RNA Tie Club in early 1955. There the idea seemed mundane. But it would prove prophetic. The paper was titled "On Degenerate Templates and the Adaptor Hypothesis." He started by saying that he wanted to expose a few thoughts to the "silent scrutiny of cold print." The first task was to dispose of all Gamow's codes by reference to the insulin and other sequences now available: "I have set out these at length, not to flog a dead horse, but to illustrate some of the simplest ways of testing a code. It is surprising how quickly, with a little thought, a scheme can be rejected. It is better to use one's head for a few minutes than a computing machine for a few days!"

Indeed, any code taken directly from DNA seemed impossible. No part of the double helix was hospitable to the hydrophobic side chains of some of the amino acids. Instead some combination of consecutive bases must tell some piece of machinery which amino acid to choose in which location. Crick then laid out his idea: that there would exist 20 different "adaptor" molecules, one for each amino acid. The job of these adaptors was to recognise some piece of code and bring the appropriate amino acid to join the sequence of a nascent protein molecule. He did not yet say explicitly that the adaptor had to be a small nucleic acid molecule, let alone RNA, but he did emphasise the likely central role of hydrogen bonding between the bases.

From now on, guided by the adaptor hypothesis, Crick was ready to abandon a "purely structural approach." In other words, despite the extraordinary fit of function to form in DNA, the next piece of the puzzle would be an arbitrary code, not a beautiful shape. DNA was an information machine, not a mechanism in itself. This explains Crick's general indifference, over the next few years, to discovering the structure of RNA. Watson found this hard to take. He wrote that he disliked the adaptor, and "we must find RNA structure before we give up and return to viscosity and bird watching."

Crick's paper ended, in keeping with his official status as the club's pessimist, on a gloomy note:

> Altogether the position is rather discouraging. Whereas on the one hand the adaptor hypothesis allows one to construct, in theory, codes of bewildering variety, which are very difficult to reject in bulk, the actual sequence data, on the other hand, gives us hardly any hint of regularity, or connectedness, and suggests that all, or almost all sequences may be allowed. In the comparative isolation of Cambridge I must confess that there are times when I have no stomach for decoding.

For "isolation," read "lack of a conversation partner." With Watson still in America, and Perutz and Kendrew preoccupied with—at last—breakthroughs in isomorphous replacement to deduce the structure of haemoglobin and myoglobin, Crick needed somebody to talk codes with. In late 1954, he found someone: Sydney Brenner. He and Brenner had first met in April 1953, when Brenner was in a party of four scientists at Oxford who came to Cambridge especially to see the double helix model. On that occasion Brenner (like most people) felt overwhelmed by Crick's loquacity and gravitated towards Watson. Then in 1954 they met again when Brenner passed through Woods Hole. In December Brenner was due to return to his native South Africa, but he went via Cambridge, where he talked to Crick about codes, just before Crick wrote the paper on adaptors. Indeed, it was Brenner who suggested the word "adaptor" and who nudged Crick towards the conclusion that the code might be degenerate—that there might be more than one way of specifying each amino acid.

Sydney Brenner's father, Morris, was a shoemaker like Harry

Crick, but a rather humbler one—an illiterate Jewish cobbler who emigrated from Lithuania to South Africa. Brought up in poverty, Sydney learned to read from the newspapers his mother used in place of a tablecloth. He got into a charity school; went to the University of Witwatersrand at age 14, qualified as a doctor at age 20; and, while waiting to become old enough to get a licence to practise, was diverted into biochemistry and won a scholarship to Oxford. Short and with a determined, almost pugnacious, manner, but brimming with jokes and anecdotes, Brenner had both the intellect and the personality to stand up to Crick's ruthless style of argument. In the Officers Training Corps at Witwatersrand University during the war, Aaron Klug and all the others were privates. Brenner was the corporal.

Brenner knew quite a lot about codes. Although he was a bacterial geneticist by training, he had been reading the Hungarian-American mathematician Johnny von Neumann on the subject of self-reproducing machines, or cellular automata. Von Neumann argued that such machines would need to store separately the information needed to make the machine and would need to have a mechanism to interpret that information—a tape and a tape reader. In effect, he abstractly described the gene, the ribosome, and the messenger. No other biologists, not even Brenner, seem to have used von Neumann's scheme to direct their own thinking in detail. But at least Brenner was thinking in terms of information, and of a decoding apparatus separate from a storage mechanism. This was probably his first and most vital influence on Crick, and it might have played a part in drawing Crick away from schemes in which DNA directed protein manufacture directly. One might legitimately ask why Crick, who talked regularly to Kreisel, had not himself read von Neumann. Kreisel recalls later mentioning von Neumann's paper to Crick, but Crick dismissed it as a mathematical abstraction.

Von Neumann himself was briefly drawn into the next of Gamow's schemes, the "combination code." Gamow had become excited about the equivalence between the number of amino acids (20) and the number of DNA base triplets (20) if the order of the letters in each triplet did not matter. Moreover, he noticed that some such triplets came in only one variety (such as AAA), whereas others came in many varieties (such as AGT, ATG, GAT, GTA, TAG, TGA). So perhaps this explained why different amino acids were—according to von Neumann's analysis—nonrandomly scarce or frequent. Crick—his intuition as infallible as ever—immediately disliked the idea. "I think it stinks," he wrote to Watson. It was hard to imagine a mechanism that would be indifferent to the order of the letters. Besides, "the supporting evidence is so weak that I cannot really take it seriously," Crick wrote to Brenner, who was now back in South Africa, adding: "My own view is that coding, etc., should be put on the shelf for a bit."

In October 1955 Crick's mother died, at age 76. Grief-stricken, Crick took to his room for three days, but then emerged fully composed. Apart from his brother Tony, now living in New Zealand, his closest remaining relatives were his uncles and aunts: Walter Crick in California; Arthur—who would die within a year—in Kent; Winifred Dickens in Northampton; and Ethel Wilkins, who had recently moved to a large house just off Madingley Road in Cambridge. Michael now moved in with his great-aunt Ethel. With a small legacy from his mother,

Crick acquired the house next door, 20 Portugal Place, knocked through a wall, and put in a sliding door. The American biochemist Alex Rich, who had worked on RNA structures with Watson and codes with Gamow, was then visiting from Caltech with his wife, Jane, and had just decided to stay longer to try to finish unravelling the structure of collagen with Crick. So the Riches moved into number 20. The Cricks employed a series of au pair girls. In exchange for food and lodging the au pair would prepare breakfast and wash nappies before going to language school in the afternoon. Briefly in 1955, in between au pairs, Linda Pauling, Peter's sister, was doing the job and living in the basement of number 20.

In May Francis and Odile went to Paris so that Crick could call on a team of scientists at the Pasteur Institute. There, for the first time, he met Jacques Monod and François Jacob, whose collaboration would come closest to rivalling his own with Watson in the pantheon of molecular biology, and whose intellectual challenges he would come to relish. Said Jacob: "We just had no idea what Crick was, so Crick was just an appendix to Watson for us—until we saw Crick, and then it was clear that Crick was not an appendix to Watson." At lunch in a café called Cosmos on Boulevard Montparnasse, the geneticist Boris Ephrussi pointed out that there was still no hard evidence that DNA base sequences defined amino acid sequences, a point Crick had to concede. But back in Cambridge, Vernon Ingram was just about to provide that evidence by linking a genetic mutation for sickle-cell anaemia to a single change of an amino acid in the haemoglobin sequence.

Having put coding on the shelf, Crick spent 1955 on lesser

scientific quests: lysozyme, collagen, and viruses. The work on lysozyme with Vernon Ingram made little progress. Collagen got Crick into trouble again. Since Randall's team at King's had been methodically working on collagen, Crick's first speculative (and wrong) paper, written in Brooklyn, had provoked furious tirades from both Wilkins and Randall. Now G. N. Ramachandran in Madras published a rather elegant triple helix structure for collagen, which Crick and Alex Rich thought they could improve on, having just built a successful three-chain model of an artificial polypeptide, polyglycine II. It was many years before they were proved right.

As for viruses, these would prove a red herring as far as coding was concerned, and they would be the last direct collaboration between Crick and Watson. At the end of June 1955, Watson arrived in Cambridge for a year's leave before going to Harvard. A few days later Crick was walking to work and bumped into Nevill Mott, Bragg's successor as head of the Cavendish. "I must introduce you to Watson," Crick said, "Since he's working in your lab." "Watson?" Mott replied. "I thought your name was Watson-Crick." Watson's plan was still to find the structure of RNA, and he seems to have hoped that another few months with Crick would recapture the first, fine careless rapture of 1953. But he also wanted to work on the structure of plant viruses, a project he had started with some success in 1952. He was now intrigued by the similar size and structure of viruses and microsomes: small round objects in cells. Microsomes seemed to contain both protein and RNA, so perhaps they had something to do with the decoding machinery and viruses would illuminate it.

Before he arrived, Watson had written to ask Crick to obtain from the virologist Roy Markham some potato virus called PVX. Crick replied that "Rosalind [Franklin] had also asked [Markham] for some. This is complicated, because I am at the moment on excellent terms with Rosalind and she shows me her results as soon as she gets them." Pinching Franklin's project again would not be a good idea, so they chose a different virus to work on. Franklin, with the help of Aaron Klug and others, was now leading the analysis of virus structure, producing a stream of papers from Birkbeck College. Watson-Crick's eventual contribution to virology was simple but central: two papers arguing—correctly, though, unfashionably—that the reason all small viruses were either rods or spheres was that they were all made up of identical protein subunits, whose job was simply to hold and protect the viral genes within. Part of their argument was that there was too little RNA in a virus to encode more than one protein subunit. But when they presented this at a small conference in London in March 1956, the argument was lost on most virologists, who still refused to admit that the ability of a virus to be infectious came purely from its RNA (plant viruses have RNA genes rather than DNA genes). Virologists had not yet caught up with Avery's proof that genes were made of nucleic acids, let alone Watson-Crick's argument that the structure of nucleic acids explained heredity.

In early April 1956, while Watson was away in Israel and Egypt, Crick travelled to Spain to present the work on viruses again to a larger meeting in Madrid. Wilkins and Franklin were among the others attending. After the conference, the Cricks travelled as tourists south to Toledo, Seville and Córdoba by train and bus, then back through France. This was the first real vacation they had taken since their honeymoon eight years before. They took with them Rosalind Franklin, who now became a firm friend of them both. She regularly probed Crick for advice these days, and in Madrid she found that she got on well with Odile—and not just because they shared a rather French preference for nice clothes and lightly cooked vegetables. When later that year Franklin had two operations to remove an ovarian tumour, it was to the Cricks in Cambridge that she went to convalesce. "Rosalind has had two mysterious operations, but is now much better," Crick wrote to Watson in November 1956, displaying a typically English aversion to medical detail. She stayed with them again the next year when her closest friend, the virologist Don Caspar, was in Cambridge. But the cancer returned, and in April 1958 she died, at age 37.

Almost unknown at the time of her death, Rosalind Franklin long afterwards became famous for the part she had played—and even more for the recognition she had been denied—in the discovery of the double helix. Many people perceived sexism. Franklin's friends, notably Klug and Crick, argued that her posthumous role as a feminist martyr would have appalled her. Like many scientists of both sexes, perhaps more than most, she suffered prejudice from rivals, some of whom were jealous and self-seeking. These rivals included not just Wilkins and Randall, but, later, the leading British virologists. Their attitude was not necessarily because she was a woman. By the standards of the time, science was remarkably welcoming to women, and King's had more senior women than most institutions. Five of the eight women in the same research group at King's at the same time as Franklin later told Horace Judson that they had experienced almost no prejudice against their sex, and several volunteered that they found Franklin aloof. The pantomime villain "Rosy" was presented in Watson's book *The Double Helix*, where she was seen largely through Wilkins's eyes as an intransigent obstacle. This portrayal infuriated many people, not least those who felt that Watson and Crick had effectively stolen, or at least obscured the debt they owed to, Franklin's data. Ironically, it was therefore Watson who lit the spark for her rehabilitation.

Watson always bore the brunt of this controversy; Crick largely kept out of it. But in 1979, Crick was briefly drawn into it by some remarks he made in an article in *The Sciences*:

> Rosalind's difficulties and her failures were mainly of her own making. Underneath her brisk manner she was oversensitive and, ironically, too determined to be scientifically sound and to avoid shortcuts. She was rather too set on succeeding all by herself and rather too stubborn to accept advice easily from others when it ran counter to her own ideas. She was proffered help but she would not take it.

Crick received some angry letters after writing this. In response to one, from the oncologist Charlotte Friend, he went even farther:

> I think she was a good experimentalist but certainly not of the first rank. . . . Her theoretical crystallography was very average. . . . What I object to is the artificial inflation of

> her reputation by women who do not fully understand her work and often did not know her personally. Rosalind would have been the first person to object to this misguided movement to make her a martyr. First-class scientists take risks. Rosalind, it seems to me, was too cautious.

Franklin's closest colleague, Aaron Klug, who saw a copy of this letter, thought it a little harsh:

> I would think by your criterion, large numbers of our colleagues could be equally found wanting. She knew very well she was no Pauling (or, for that matter as it turned out, no Crick). What distinguishes her from the select few was that she was not highly imaginative; but how many scientists are?

Besides, says Klug today, who was the person that most people considered "not sound" in 1953? Not the careful Franklin but the flashy Crick.

Chapter Seven

Brenner

SOON AFTER HE GOT BACK from Spain in April 1956, Crick wrote to Sydney Brenner in South Africa to tell him of a new coding idea. "Re coding: Leslie [Orgel], John Griffith & I have deduced the magic 20, using a code having 3 bases to 1 amino acid." This was the famous "comma-free code," an idea so elegant that it had to be right—but it was wrong. It has been called possibly the most beautiful wrong idea in science, or the moment when Crick beat God at his own game. Yet Crick never really believed in it; he was fully aware that it was just speculative.

It all started with a talk by Leslie Orgel in February about the problem of "punctuation" in the genetic code. Orgel, like Brenner, had been in the car full of scientists who travelled from Oxford to see the double helix in 1953. After leaving his post at Oxford he had gone to work in Pauling's lab and had worked with Watson on RNA, getting drawn into Gamow's coding schemes. Now he was a lecturer in inorganic chemistry at Cambridge, but in Brenner's absence he was a good foil for Crick's ideas, and his talk started Crick thinking again.

By now they were assuming that DNA carried a triplet code: three bases for each amino acid, for the simple reason that one- and two-letter codes would allow only four and 16 combinations respectively. They thought all overlapping codes, like the one that had excited Gamow, could be discarded because such codes demanded forbidden combinations of adjacent amino acids and none of the proteins yet sequenced seemed to have such constraints. A nonoverlapping code raised the problem of how each adaptor bearing an amino acid "knew" where a triplet began and ended. Crick was not yet thinking of a decoder moving along the DNA; rather, he was thinking of the amino acids as simultaneously assembling themselves at the right places on the sequence, albeit with adaptors to bring them there. Accordingly, he ruled out something as stupid as starting at one end and counting in threes, which would anyway be vulnerable to phase shifts if a letter was skipped. He wondered if there might be a way for all misreadings that spanned more than one "word" to be automatically nonsensical. He mentioned the idea to Orgel, who immediately saw that this could produce no more than 20 "sense words." Twenty was the magic number of amino acids.

The reasoning was simple. Start with all 64 triplets. First discard the four triplets AAA, CCC, GGG, and TTT—one could not be next to another of the same kind without producing a spurious triplet in the wrong place overlapping the two. That leaves 60. Now imagine every cyclic combination of the same three letters (such as ACT, TAC, and CTA) and choose only one from each, because if ACT is legal, then the sequence ACTACT could be read as having CTA or TAC in the middle: so CTA and TAC must be illegal. Choosing only one from each cyclic combination of three letters means dividing 60 by three, which leaves 20.

But this did not prove that there must be 20 unambiguous triplets; it proved only that there could not be more than 20. The problem was to prove that from three-letter combinations of four letters, you could find a set of 20 triplets such that when two triplets were next to each other, no triplet in the overlap zone made sense. Crick went to bed with a cold and tried to work out a set of 20 triplets that had this property, but he could get only to 17. Orgel mentioned the problem to John Griffith, the mathematically trained chemist who had helped Crick with base pairing in 1952, and Griffith quickly found a set of 20. Indeed, he soon calculated that there were at least 288 solutions, all of which gave 20 triplets.

The beauty of such a code would be that making a mistake in it was remarkably difficult. It could work by having adaptors that fitted the "legal" but not the "illegal" triplets. At this point, excitement began to rise. Using pure logic and no experiments, Crick and the others had found a way of writing a code for a 20-letter alphabet using three-letter words in a four-letter alphabet: precisely the problem to be solved. But the very purity of their method caused Crick to hesitate. They had no other evidence at all for the comma-free code. It was a castle in the air, a fantasy for which there was no evidence. Mathematicians like such things—indeed, the mathematician Solomon Golomb would shortly go on to work out the full of set of all possible triplet codes having such a property. But Crick was an empiricist interested in the real world. He refused to get carried away.

Nonetheless, news spread that Crick, Griffith, and Orgel had cracked the genetic code. They wrote the idea up as a note for the RNA Tie Club; then, to satisfy the demand from people wishing to quote a reference, they published it in the prestigious *Proceedings of the [American] National Academy of Sciences*, where it appeared in 1957, full of suitable disclaimers. "The arguments and assumptions which we have had to employ to deduce this code are too precarious for us to feel much confidence in it on purely theoretical grounds." Soon it was reported almost as fact by *Scientific American*, and Ruth Moore put it in her book *The Coil of Life*—published long after Crick had ceased to believe in the idea. "I was in the embarrassing position," he later wrote, "sometimes of finding that people believed it more than I did."

At the end of April 1956 Crick went to the United States for the summer; he would be there till mid-August, leaving Odile and the children to spend the summer with her mother in Norfolk. He celebrated his fortieth birthday in June in Alex Rich's lab in Bethesda, Maryland, outside Washington, where he was working on the structure of collagen and on an artificial RNA made of a string of adenines. While Crick was in Bethesda, he made a valiant but vain attempt to learn from Rich how to drive a car, an experience that Rich said "diminished both our egos."

Moving on to Baltimore afterwards to speak at a meeting on the chemical basis of heredity, Crick found that he and Watson had been given the presidential suite at the Baltimore Hotel, the first sign that the double helix had begun to bring them real renown. Chargaff, speaking before Crick, brought him back to earth by caustically denigrating the attention being paid to DNA. In Madison, Wisconsin, where Crick had been invited to give a lecture for "a small group of about 15," he attracted an audience of 200. In mid-July in Ann Arbor, Michigan, where he was to deliver three lectures, Crick wrote two long letters—one to Watson and one to Brenner—summarising what he had learnt about the current state of genetics. The letter to Watson was a good example of how quickly Crick buried personal disagreements. A few weeks earlier, in the tone of an older brother, he had furiously scolded Watson for failing to finish his paper on viruses for the London symposium held earlier in the year, and thus holding up publication. "I need hardly say I am extremely displeased about this. I also do not like having to make excuses when I go to conferences about your absence. This sort of thing was, perhaps, excusable when you were in your teens, but not when you are in your late twenties." Now, all that forgotten, Crick was full of news and ideas, as he groped towards a theory of how protein is synthesised. He suggested as a "postulate" that "the microsomal particles are the only (cytoplasmic) site of protein synthesis"—an accurate guess.

Crick's letter to Brenner ended, "I can hardly wait for you to arrive. Is there no chance of your arriving a month or two earlier?" Crick had for some time been eager to get Brenner a job in Cambridge, and the previous autumn an opportunity had arisen when John Kendrew's wife left him for Hugh Huxley, necessitating Huxley's sudden departure. Crick needed somebody to do experiments for him, and he also needed somebody to bounce ideas off, somebody who enjoyed talking as much as he did. Brenner was to be recruited to fill both roles.

Before Brenner arrived at the end of the year, though, there was a real breakthrough. The adaptor, till now an entirely theoretical concept, became incarnate. It was a case of simultaneous discovery by three labs at once, though Paul Zamecnik and Mahlon Hoagland at Massachusetts General Hospital probably had the clearest claim. Zamecnik had developed a way to show that microsomes, extracted from cells and held in a test tube, could assemble proteins from radioactively labelled amino acids; and now Hoagland found that before being incorporated into protein, each amino acid spent some time attached to a small, soluble RNA molecule. Hoagland, who had never heard of the adaptor and was not part of the "coding cabal," was crestfallen when told by Watson at the end of 1956 that his discovery had in effect been interpreted before he had even made it. He compared himself to an explorer slashing and sweating his way through a jungle, "rewarded at last by a vision of a beautiful temple—looking up to see Francis, on gossamer wings of theory, gleefully pointing it out to us!" Yet Crick refused to believe that this soluble (or transfer) RNA was indeed the adaptor: it was much bigger than he had expected. Only later, as it gradually became clear that each kind of amino acid had a specific individual kind of transfer RNA, did the precise fit between the theoretical adaptor and the actual transfer-RNA become obvious.

Crick flew back to Britain in mid-August 1956. Sydney Brenner arrived in December and moved into Portugal Place while he hunted for a house. He also moved into Crick's office in the Austin wing. For the next 20 years he and Crick would share a book-filled office and talk virtually every day, starting after coffee in mid-morning and sometimes continuing through lunch at the Eagle or the Friar House, and on to the all-important cup of tea accompanied by several sweet biscuits in mid-afternoon. (Those biscuits were a lifelong addiction of Crick's.) The blackboard was the focus of most discussions, accumulating a jumble of words and diagrams. Anand Sarabhai, who would prove the colinearity of gene and protein, recalls that the blackboard continually changed its appearance, as many times during the day as new theories, speculations, and facts emerged. The dialogue between Brenner and Crick was a conversation that developed its own rules. There was no shame in floating a stupid idea; but no umbrage was to be taken if the other person said it was stupid. Anyone else from the lab could walk in and interrupt if the door was open, but strangers were directed to see the secretary. Like Watson, Brenner knew a lot more biology than Crick. Brenner found Crick an "incredible cross-examiner" who always challenged him on how to test an idea with a real experiment. Crick's other "work" consisted mostly of reading scientific papers. He was a ravenous consumer of others' results, from even the most obscure publications, and he had formidable powers of concentration. Aaron Klug once asked why Crick was wasting time on an obviously useless paper. "There might be a clue in it."

Brenner's job, in between arguments, was to set up a "phage" laboratory for which Crick begged, stole, and borrowed space, equipment, and money. A phage—or, to give it its proper name, bacteriophage—is a virus that attacks a bacterium, subverting the machinery of the bacterium to make more virus. Its presence was easily detected as clear plaque of dead bacteria on an opaque "lawn" of bacteria growing on an agar plate. Mutant versions of the phage failed to kill the bacteria and so produced no plaques or small plaques. What Brenner wanted to do first was prove the "colinearity" of a gene with a protein—to prove that the sequence of bases in a gene precisely determined the sequence of amino acids in a protein.

The gene he would do it with, in phage, was known as rII. At Purdue University, Seymour Benzer—an extraordinary experimentalist from a humble background in Brooklyn who had already achieved distinction in electronics before turning to genes—had discovered that by crossing hundreds of different mutant phages with each other, each one having a mutation in this same gene, he could effectively create a map of the structure of the rII gene down to the level of a single base pair. This was in effect destroying the old concept of a gene as an indivisible entity: it was "splitting the gene." Benzer, who would go on to discover the genes behind memory and mating in fruit flies, came to Cambridge in the autumn of 1957 to join Brenner. The latter was startled by Benzer's nocturnal habits and low tolerance for even the slightest cold weather. They set out to find a chemical that would mutate one base to another, see which amino acid the virus then substituted, and so decipher the genetic code. At least that was the plan.

Crick now sat down and thought through from scratch the whole problem of translating DNA to protein, sifting the evidence, discarding misleading hints, and distilling out the truth. The result was probably his most remarkable paper, which he delivered in 1958 to the Society of Experimental Biology at its meeting in Canterbury. Called simply "On Protein Synthesis," it defined the field. A little like Newton's *Principia*, or Wittgenstein's *Tractatus*, Crick's paper makes a set of bold assertions that depend on each other. The function of genes is to make proteins. There are 20 kinds of amino acids in proteins, and all 20 occur in nearly all proteins, whatever the species of organism. Proteins have defined and fixed amino acid sequences. The folding up of a protein is simply a function of the order of the amino acids. That order is determined by the order of the bases in a gene. Proteins are made mainly in the cytoplasm, not the nucleus. They are made at "microsomal particles" (soon to be known as ribosomes). Specific adaptors made of nucleic acid bring the amino acids to the site. The code used is written in nonoverlapping triplets of bases.

All these propositions were guesses, and all are correct. Crick even prophesied the use of protein and DNA sequences in genealogy and taxonomy: "Vast amounts of evolutionary information may be hidden away within them." From the mass of confusing results that had been pouring into the literature, he had somehow drawn nearly all the right conclusions and had been distracted by almost none of the red herrings. This intuitive talent, which is evident throughout his career, was what made him so valuable to his colleagues and is what they still find hardest to explain. Crick did not get everything right. In "On Protein Synthesis," he suggested that the RNA found in microsomes was the template for protein synthesis. This error would take two years and a flash of inspiration from Brenner to correct.

The most remarkable part of the paper is the two general principles that Crick then draws from all the evidence:

> My own thinking is based on two general principles, which I shall call the Sequence Hypothesis and the Central Dogma. The direct evidence for them both is

negligible, but I have found them to be of great help in getting to grips with these very complex problems. I present them here in the hope that others can make similar use of them. Their speculative nature is emphasised by their names. It is an instructive exercise to attempt to build a useful theory without using them. One generally ends in the wilderness.

The "sequence hypothesis" is that a sequence of DNA bases determines a sequence of amino acids and that nothing else is then needed to tell a protein how to fold. This was still heresy to most biochemists, but it was becoming orthodox in Crick's circle. It is the fundamental surprise of molecular biology. At the time it was still an assumption, albeit an increasingly plausible one. The "central dogma," on the other hand, would become controversial and even notorious. In its first formulation here, it states:

> Once "information" has passed into protein, it cannot get out again. In more detail, the transfer of information from nucleic acid to nucleic acid, or from nucleic acid to protein may be possible, but transfer from protein to protein, or from protein to nucleic acid is impossible.

Crick had used the term "central dogma" a few months before in *Scientific American*, where he made it clear that he was forbidding proteins to copy themselves or to change their own nucleic acid recipes. The central dogma would cause a small disagreement with Watson that rumbled on till Crick's death. It has often been rendered, in simple form, as "DNA makes RNA makes protein." This was a statement Watson rightly claimed he had first made, in a letter to Delbrück in 1952. Crick's emphasis was on the fact that proteins were recipients but not donors of sequence information.

As the historian Robert Olby would point out later, Crick was trying to kill a belief that had so far refused to die: the belief that the relationship between DNA and proteins was reciprocal, that DNA determined protein sequences but proteins also determined DNA sequences, and that "genes" were therefore a combination of both. This was true in a biochemical sense, but it was entirely false in the sense of information. The information required to assemble a protein sequence lay in a DNA sequence; the information required to assemble a DNA sequence also lay in a DNA sequence. Crick's use of the word "dogma" would cause much trouble in the years ahead, especially with Barry Commoner, an unreconstructed partisan of protein who spent his career clutching at every straw that suggested the double helix was an error until Crick called him "wilfully obtuse." As late as 2002, in *Harper's* magazine, Commoner would assert, with a strange leap of logic, that the Human Genome Project had disproved the central dogma by finding that more than one protein could be made from different sections of one gene by alternative splicing. As is clear from the original context, Crick meant by the word "dogma" that he knew there was as yet no proof of this. It was, therefore, more like a bold speculation. Despite many attempts to topple it, the central dogma remains true: base sequences in DNA determine amino acid sequences in protein, but not vice versa.

The next two years saw little progress. The code remained elusive and the phages intransigent. Almost 150 pages of handwritten coding schemes accumulated in Crick's files. At least there was good news from Caltech, where Matthew Meselson and Franklin Stahl had ingeniously provided the first independent proof of the double helix, by showing that when genes are duplicated in dividing cells, DNA is "semiconservatively replicated" with each strand of a double helix serving as the template for a new strand—just as had been predicted in 1953. In March 1958, Crick wrote to Watson at Harvard: "Sydney has had some newish ideas about coding, which we will tell you about when they are a little tidier. . . . Gabrielle and I have had German measles, Jacqueline has mumps, Odile is covered with spots of unknown origin but otherwise we are all well." (This reminds one of the old joke: "Apart from that, Mrs Lincoln, how did you enjoy the play?") In May Crick travelled to Paris and gave a talk, in French, at the Pasteur Institute. Not daring to trust to spontaneity in French, he had written the whole talk out with Odile's help and had removed the jokes because they looked bad on the page. It was not a great success. That spring he also moved out of the Austin wing a short distance to a new office in "the hut," an unprepossessing single-storey brick shed with a pitched roof, which still stands in a courtyard of the New Museums Site. He continued to share a room with Brenner.

In March 1959 Crick was elected a fellow of the Royal Society, having been put forward by Perutz and Bragg. The latter had written in his recommendation, "Crick has the most lively, intelligent and speculative mind," but had added waspishly, "I have never been quite clear how much was Watson and how much

Crick, because Crick did all the talking." At this time Crick was at Harvard for a term as a visiting professor, staying with the Riches, but he travelled so much that he could hardly pause to consider Harvard's offer of a permanent job. In April he and Odile took the children to New Orleans and then to Gatlinburg in the Great Smoky Mountains of Tennessee for a brief family holiday, before a spell in Berkeley. In June, suffering from sunburn, he gave a talk at a symposium at the Brookhaven National Laboratory on Long Island in which he frankly admitted that as far as the code was concerned he was now floundering. He said that the coding problem had gone from its vague phase through its optimistic phase and was now in its confused phase. The previous summer two Russians had published an analysis of the base composition of DNA and RNA in bacteria. The DNA varied enormously, some species having five times as high a proportion of guanine and cytosine as others; but the RNA was always much the same. Perhaps the code was not universal among all species. Perhaps it was degenerate, with many different ways of spelling the same amino acid. Perhaps it was full of junk messages that were unrelated to coding. Crick mentioned all these possibilities but admitted that he was discouraged; and in the discussion afterwards he had to fend off suggestions that maybe genes were really made of RNA—or even pure sugar sequences. "The whole business of the code was a complete mess," he later said. "We were completely lost, you see. Didn't know where to turn. Nothing fitted."

In 1957 he had he applied to succeed the great evolutionist Sir Ronald Fisher, who had reconciled the theories of Mendel and Darwin, as Arthur Balfour Professor of Genetics at Cambridge. Fisher encouraged his candidacy, but Cyril Darlington, who was on the committee making the appointment, ensured that Crick was turned down in favour of a population geneticist, John Thoday—more evidence, apparently, that geneticists were happier with abstract than real genes. Darlington, a senior geneticist at Oxford with a somewhat cantankerous manner, was one of those who still refused to believe that DNA directed the manufacture of protein, preferring to regard the relationship between them as equal and reciprocal—precisely the idea that the central dogma rebutted.

This is a reminder that, scientifically, things were not now going Crick's way. The theory spelled out so logically in "On Protein Synthesis" was a minority view. And new results were increasingly hard to square with it. In September 1959 Crick went to Copenhagen for a meeting organised by Watson's former colleague Ole Maaloe. There, the hot news was from Paris, where Jacques Monod and François Jacob had done a beautiful series of experiments that simply did not fit Crick's ideas. Monod, six years Crick's senior, was an almost absurdly gifted man. Sailor, rock climber, cellist, orchestra conductor, communist, French resistance fighter—he had many distractions to keep him from the laboratory bench until about age 40. But then in the mid-1950s he proved that a bacterium could rapidly switch on the manufacture of one of its proteins in response to the presence of lactose sugar, the first evidence of gene switches. Jacob, whose body was full of shrapnel received when fighting with the Free French Army in 1944, had then developed a brilliant technique for interrupting the "conjugation" of bacteria as they transferred genes in sequence from one to another, thereby mapping the genes along the chromosome. Now, with Arthur Pardee, a visiting Californian, Jacob and Monod had shown that when one bacterium transfers a gene to another, the latter begins to produce the protein specified by that gene within three minutes—much too quickly for the manufacture of new ribosomes. This "PaJaMo" experiment (named after Pardee, Jacob, and Monod) was incompatible with Crick's assumption in "On Protein Synthesis" that each ribosome carried an RNA copy of a gene and made a particular protein. So Crick was unconvinced by the experiment: if the facts don't fit the theory, he said, first question the facts.

In April 1960, Jacob was in London for a conference and travelled to Cambridge for the Easter weekend. On Good Friday, 15 April, the labs being closed, Crick, Brenner, and others gathered in Brenner's room in King's, to listen to Jacob again recount his baffling tale. He found a sceptical audience. "Francis and Sydney made me take a veritable examination! With questions, criticisms, comments. A pack of hounds racing around me nipping at my heels." Jacob held his ground. He described new evidence that very soon after a gene was deliberately destroyed by the decay of radioactive phosphorus, protein manufacture ceased. Then suddenly Brenner let out a "yelp." He began talking fast. Crick began talking back just as fast. Everybody else in the room watched in amazement. Brenner had seen the answer, and Crick had seen him see it. The ribosome did not contain the recipe for the protein; it was a tape reader. It could make any protein so long as it was fed the right tape of "messenger" RNA. And that messenger had been discovered four years before when Eliot Volkin and Lazarus Astrachan had found a kind of unstable, free RNA whose composition mirrored that of phage DNA. Volkin and Astrachan thought they had found an intermediate used in making more DNA, but in fact they had found the RNA copy of a gene that was read by the ribosomes and translated into proteins. Brenner could have kicked himself, because this was exactly von Neumann's theory of self-reproducing machines: the tape and the tape reader. Crick later wrote of the moment of insight: "It was so memorable that I can recall just where Sydney, François and I were sitting in the room when it happened."

Brenner and Jacob immediately planned an experiment, which they carried out in California that summer, to prove the existence of the RNA messenger. (The experiment would lead, temporarily, to some bad blood between Brenner and Watson, whose Harvard lab, six weeks prior to Jacob's visit to Crick and Brenner, had completed phage experiments showing that ribosomal RNA did not specify amino acid sequences.) Crick sat down, probably that afternoon, and wrote a paper about the new understanding, though he never published it. That evening there was a party at Crick's house. But the wine and women were not as distracting as usual. The scientists stood talking to each other about messenger RNA. Its "discovery" that day had been sudden but was overdue. Unlike the adaptor, which Crick had predicted a year before it was found, the messenger had to wait four years after being found (by Volkin and Astrachan) before it was recognised. Because a ribosome made proteins and because it contained much RNA, everybody had been fixated by the notion that its RNA must consist of copies of genes. Now it was clear that the protein-directing RNA came from outside and that the ribosome's own RNA was merely part of its makeup (though that RNA did also originate in special genes). This realisation got rid of the whole body of awkward facts that had been blocking progress on coding: facts which seemed to show that ribosomal RNA never varied much from cell to cell, from species to species, or from protein to protein. The attack on coding could now resume.

Chapter Eight

Triplets and Chapels

IN AUGUST 1960 Crick received news that he, Wilkins, and Watson were to be jointly awarded the Lasker Prize by the American Public Health Association for the double helix. With it came $2,500 each and a heavy statuette of the Winged Victory. Perhaps more important, everybody knew that the Lasker Prize was a frequent harbinger of the Nobel Prize. The Lasker was soon followed by the Prix Charles Léopold Meyer of the French Academy of Sciences, and in 1962 by the Award of Merit of the Gairdner Foundation in Canada.

Some time in 1961 Crick had a 3-foot metal helix made in the lab's workshop. It was painted gold and erected above the door of 19 Portugal Place, which he now renamed the "Golden Helix." It was a single, not a double, helix, indicating his pride in his first breakthrough in helical theory. Life at the Golden Helix was settling into a rhythm. Odile was painting and making pots in the third-floor studio. Gabrielle and Jacqueline found their father "not a hands-on, bedtime-story, teach-you-how-to-ridea-bicycle kind of dad," but he was a kindly presence. His efforts to teach them science, using items from the fruit bowl on the dining table to stand in for planets or particles, sometimes palled. The family only very rarely went to concerts or to the cinema. When Michael asked his father why they did not go to films more often, Crick pointed out that watching neurotic people on-screen was no better than doing so in real life. In the early years the family had no radio, no television, few magazines, and no daily newspaper. One of Michael's chores was to buy a copy of the *Observer* every Sunday morning for his father to read in the bath.

But there were frequent parties, renowned for the pulchritude of the female guests and for the free-flowing punch. On one occasion the guests were asked to come dressed as "beachcombers or missionaries." At another time, on Friday, 1 June 1962 at nine in the evening, the occasion was a "studio party" at the Golden Helix, for which the guests were dressed as "artists, models, or dancing girls." A sketch of a nude, by Odile, embellished the invitation. When each guest arrived, he or she was handed a sketch pad and pencil and was encouraged to draw a nude model who was posing on a couch under a window in the studio. In staid Cambridge, this caused quite a sensation.

In February 1961, Crick took to the lab himself. He had an idea he wanted to test, and he had grown tired of waiting for Brenner to take it seriously. He taught himself phage-crossing techniques, picking tiny samples from plaques, crossing them with others on new bacterial lawns, and incubating the resulting plates for a few hours at 37 degrees Celsius. He was predictably clumsy and predictably argumentative with technicians about the reasons for doing things in certain ways. But he was determined to learn. He used two strains of bacteria, of which one was immune to phages in which a particular gene was broken, and the other was susceptible. This way he could find mutations in the viral gene. The question that interested Crick was how one mutation suppressed another. Brenner and Alice Orgel had found that some chemicals that caused mutations in the viruses could also "cure" or suppress a mutation, but not if it had been caused by a different chemical. So a yellow acridine dye called proflavin could cause a mutation and then cause a reversion to normal activity. One Saturday in November 1960 Brenner had had the idea—predictably, it came to him while he was eating in the Eagle with Crick—that this was because proflavin, instead of substituting a letter in the code, inserted or deleted a base in the sequence.

Crick had a hypothesis for how two mutations suppressed each other. He believed that the messenger RNA twiddled itself into a loose double helical loop to show its message to the ribosomal machinery and that an inserted letter could be geometrically corrected by another insertion at the opposite side of the twiddle. He therefore expected to map mutations that suppressed each other to distant spots on the viral chromosome.

Instead, when he eventually mapped three suppressor mutations, in each case the suppressor was very close to the mutation it suppressed. In May he conceived an idea for a new set of experiments. He took one of these mutations, which was called P13 but which he now renamed FC_0, and looked for other mutations that suppressed it, then for suppressors of the suppressors, and then for suppressors of the suppressors of the suppressors.

In each case, the new mutation partly, but not wholly, mended the effect of the first mutation. By midsummer he had found more than 25 pairs of mutations and their suppressors.

During these weeks, the comforting routine of the experiments—two crosses a day and hours of incubation interspersed with bouts of plaque-picking—made Crick surprisingly content. He worked through most weekends but took Mondays off to let the technicians catch up with the dishwashing and preparation. The experiments were done in a sealed-off corridor in the zoology museum, which had previously housed a whale skeleton and which the professor of zoology had unwisely allowed Crick to colonise. One time, as recounted in Crick's memoirs, a "glamorous friend"—from Odile's artistic set—came into the lab late in the evening and ran her hands through his hair, saying "Come to a party." In vain. He was obsessed.

None the less, with the Lasker money, that summer Odile persuaded him to go on a real vacation. It started at the end of June with a scientific meeting at a hotel at Col de Voza halfway up Mont Blanc, and ended with the International Biochemical Congress in Moscow, but in between there was a family holiday for all of July and some of August. Through a friend they rented a villa in Tangier, on a rocky promontory at the very tip of the peninsula that divides the Atlantic from the Mediterranean. It came with a live-in servant called Mohamed and extra help by the day. While Odile, the aristocratic German au pair Eleonore Broemser von Rüdesheim (together with her boyfriend, the future architect Sumet Jumsai), and the children played on the beach or shopped in the souk, Crick read scientific papers on the terrace in the shade of the palm trees. After he left for Moscow in

August, the family stayed on for a week longer. What happened in Moscow will be told in chapter 9.

When Crick returned, he started testing so-called uncles and aunts: would a suppressor of one suppressor also suppress another suppressor of the same "rank"? Yes. The suppressors divided neatly into two kinds, which he called plus and minus. Any plus could correct any minus and vice versa; but a plus could not correct a plus, and a minus could not correct a minus. He knew by now that the twiddle theory was wrong, and he knew what was really going on. Some of the mutations caused by acridine dyes were insertions of an extra base; others were deletions of an existing base. An insertion could suppress a deletion or vice versa by getting the message back on track after a short burst of nonsense—by correcting a frame shift. Each "rank" was therefore all deletions or all insertions. Next, Brenner suggested a way of making triple mutants, each with three insertions or three deletions near the left-hand end of the gene. If the code was written in triplets, adding or taking out one or two letters would mess up the message, but adding or deleting three letters would get the message back in phase and restore the gene. By this time the experiments were being carried on mostly by the lab technician Leslie Barnett, who would go on to be senior tutor of Clare Hall, where a building has since been named in her honour. One evening after dinner she and Crick took the first triple crosses out of the incubator, and there were the telltale plaques to prove that the triple mutants were effectively normal. "Do you realise," Crick said to Barnett, "that you and I are the only people in the world who know it's a triplet code?"

The experiment also proved that the code must be read from

a set starting point, counting in threes from there. This was the faintly idiotic way of doing things that Crick had hoped to reject when he was designing the comma-free code. But it was the way Mother Nature had chosen. In addition, Crick and Barnett found that they had at last exploded the numerological fascination with getting from 64 to 20, because the sheer number of suppressors they had identified ruled out the idea of nonsense triplets implied by the comma-free code. Every triplet stood for an amino acid, even if the "wrong" one. So the code must be degenerate—like Michael's home-made cipher in 1948—with each amino acid encoded by several different triplets. There was no magic, clever way of getting 20 out of 64—just a lot of redundancy.

Those who knew Crick as a theoretician suspected that others had done the experiments while he had cheered them on from the sidelines. But in this case they were wrong. Crick had neatly and diligently ploughed his way through the experiments himself, designing each cross and carefully scoring the result. The paper on the triplet code, "General Nature of the Genetic Code for Proteins," was published in *Nature* on the penultimate day of 1961, with Barnett, Brenner, and the young physicist Richard Watts-Tobin as coauthors. It remains a landmark in the history of molecular biology, and it is very different from most of Crick's other papers because it recounts his own experiments. Unusually for the time, it received widespread coverage in the newspapers on New Year's Eve. "Scientists have cracked the code of life," said the *Sunday Times*. "A major advance in the unravelling of the secret of life appears imminent," said the *Observer*, before leaping , as newspapers will, far into the future:

"The prospect of breeding, at will, geniuses and monsters, creatures resistant to all diseases or with different instincts, is still very remote. It may never materialise—but it is no longer quite science fiction." These reports were stimulated not so much by Crick's own results as by his last two paragraphs, where he related a "startling" announcement that had been made at the Biochemical Congress in Moscow in August. Even before he had proved that the code of life is read in threes from a set starting point, the first triplet had already been deciphered. As a result, said Crick, "If the coding ratio is indeed 3, as our results suggest, and if the code is the same throughout Nature, then the genetic code may well be solved within a year."

Fame of a different kind had come Crick's way that autumn. The year before, he had become a founding fellow of Churchill College. Established to honour Winston Churchill, on the initiative of Lord Cherwell, Churchill's wartime science adviser, this college was one of several attempts by the British to imitate MIT as a specifically scientific college and thus remedy a perceived national shortage of scientists and engineers. Crick had initially refused a fellowship, because he had heard the college intended to build a chapel. This had not been part of the original plan, but under pressure from pious folk, the trustees of the college had conceded that they might build a chapel if funds became available, though Sir Winston—no great churchgoer himself—was lukewarm, saying, "A quiet room will do." Sir Edward Bullard, the professor of geophysics and a friend of Crick's from Admiralty days, who was already a fellow, came around to persuade Crick to change his mind. The chapel fund had only 10 guineas in it, Bullard said, donated by Reverend

Hugh Montefiore, dean of Caius College. It would probably never be built. So Crick became a fellow.

He had reckoned without Montefiore's cunning. Montefiore started looking for wealthy patrons who could fund a chapel, and he lit on Timothy Beaumont, an ordinand, future Liberal politician, and future Green Party peer, who had just inherited a large fortune. Beaumont donated the entire cost of the chapel, £30,000. Its foundations were dug before the fellows demanded a say in the matter. The issue came to a head in the summer of 1961, when Crick was in Tangier. A meeting between the fellows and trustees was called, but before it was held in September 1961, Crick simply resigned, feeling that he had been misled when he agreed to join the year before. He later regretted acting so precipitately and wished that he had stayed to argue the point.

He sent a short note to Sir Winston Churchill explaining his resignation as a fellow. He received the following reply:

> I was sorry to learn that you have resigned from Churchill College, and am puzzled by your reason. The money for the chapel was provided specifically for that purpose by Mr Beaumont and not taken from the general college funds. A chapel, whatever one's views on religion, is an amenity which many of those who live in the College may enjoy, and none need enter it unless they wish.

Crick replied, from the Golden Helix, on 12 October, with an outrageous proposition:

To make my position a little clearer I enclose a cheque for ten guineas to open the Churchill College Hetairae [courtesans] fund. My hope is that it will eventually be possible to build permanent accommodation within the College, to house a carefully chosen selection of young ladies in the charge of a suitable Madam who, once the institution has become traditional, will doubtless be provided, without offence, with dining rights at the High Table.

Such a building will, I feel confident, be an amenity which many who live in Cambridge will enjoy very much, and yet the institution need not be compulsory and none need enter it unless they wish. Moreover, it would be open (conscience permitting) not merely to members of the Church of England, but also to Catholics, Non-Conformists, Jews, Muslims, Hindus, Zen Buddhists and even atheists and agnostics such as myself.

[The trustees may] feel my offer of ten guineas to be a joke in rather poor taste. But that is exactly my view of the proposal of the Trustees to build a chapel, after the middle of the 20th century, in a new college and in particular one with a special emphasis on science. Naturally some members of the college will be Christian, at least for the next decade or so, but I do not see why the college should tacitly endorse their beliefs by providing them with special facilities. The churches in town, it has been said, are half empty. Let them go there. It will be no further than they have to go to their lectures.

> Even a joke in poor taste can be enjoyed, but I regret that my enjoyment of it has entailed my resignation from the college, which bears your illustrious name.

Churchill did not reply. Pencilled on the corner of the letter are the words "cheque returned with comps."

The dispute over the chapel did not end there. A group of the remaining fellows demanded that the chapel be a meditation room, available for Christian services but not dedicated to them. There should be no permanent cross in it. Montefiore denounced this stance as anti-Christian, and Beaumont refused either to agree to it or to withdraw his benefaction. Throughout the winter, Cambridge was agog. Eventually, however, the issue was defused by a messy compromise in which the chapel was built outside the college grounds. A witty rumour spread that Crick had been offered a fellowship at King's but would accept it only if King's demolished its exquisite fifteenth-century chapel. In fact, a few years later, Crick became an honorary fellow of Churchill—to "let bygones be bygones."

Though the reference to "the next decade or so" was made partly to tease, Crick genuinely seems to have begun thinking that religion was dying. Two years later he donated £100 to the Cambridge Humanist Society for a competition to choose the best essay on "What Can Be Done with the College Chapels?" The winning entry (the judges included the novelist E. M. Forster) suggested that they be made into swimming pools. In response there was a rumour that the college chaplains were offering £100 for an essay on "What Can Be Done with Dr. Crick?" Kendrew sent cuttings of both announcements to

Watson at Harvard, suggesting that he submit an entry to each contest. In 1966 Crick wrote an essay, "Why I Am a Humanist," for *Varsity* magazine. "In recent years molecular biology has practically obliterated the distinction between the living and the non-living," he wrote. "The simple fables of the religions of the world have come to seem like tales told to children." In response to a letter to *Varsity* from the biologist W. H. Thorpe, he put it more pithily: "I should perhaps emphasise this point, since it is good manners to pretend the opposite. I do not respect Christian beliefs. I think they are ridiculous."

Crick had for some years refused to attend weddings, funerals, or baptisms in churches, going only to the parties afterwards. But he sensed that new forms of ritual would have to be invented to replace religious ceremonies, if people were to join him in humanism. He even devised a new grace to be said before college meals ("Let us remember today our fellow men who have laboured that we may eat"), but he quickly ran into the paradox of humanism—that the more formal, ritualised, or intolerant it becomes, the more like a religion it seems. Earnest atheists can be almost as little fun as earnest believers.

Chapter Nine

The Prize

PART OF CRICK'S MOTIVATION for suddenly resigning from Churchill was probably his excitement because the code was about to yield, and the last thing he needed was a distracting argument with vicars. The news that startled Crick in Moscow in August 1961 came from Marshall Nirenberg, who till then had been unknown to the self-selected coding elite. Three days into the International Biochemical Congress at Moscow University, Nirenberg gave a 15-minute talk in a classroom. The talk was sparsely attended, but Matt Meselson heard it and went straight to Crick, who quickly added Nirenberg to a session he was chairing at the end of the meeting so that Nirenberg could repeat it. What Nirenberg announced was that in his little-known laboratory at the National Institutes of Health in the suburbs of Washington he had perfected the techniques, first invented by Paul Zamecnik, of getting ribosomes to make proteins in a test tube, and had quickly realised that to do so he needed to add stretches of RNA. A German colleague on a NATO scholarship, Heinrich Matthaei, had then systematically tested different kinds of RNA to find out what proteins they caused to be made. At three o'clock in the morning on 27 May 1961, Matthaei had tried recently synthesised artificial RNA made entirely of uracil: poly-U, as it was known. (Uracil is RNA's version of thymine.) The ribosomes made pure polyphenylalanine. That meant, since Crick now knew for sure that the code was triplet, that the first "word" had been cracked: UUU denoted phenylalanine.

While Nirenberg was in Moscow, Matthaei called him and reported having repeated the trick with cytosine and cracked a second word. Poly-C seemed to make polyproline. Ironically, this technique—feeding synthetic RNAs to ribosomes and seeing what the ribosomes made from them—was one that Brenner and Crick had scorned as impractical. But it was the logical next step from the discovery of the messenger. Nirenberg and Matthaei's triumph was the first hint that the new molecular biologists were becoming a bit like other scientists: it took obscure heretics to make breakthroughs, because the elite clung to old orthodoxies.

Crick arrived back from Moscow, abandoned the dispute over the chapel at Churchill College, and set out to join the heretics. He quickly started a collaboration with Marianne Grunberg-Manago, a French biochemist who had discovered the enzyme that made synthetic RNAs possible; and he redirected a postdoctoral visitor, Jim Ofengand, to learning Nirenberg's cell-free system. By October, when Mark Bretscher joined the team as a PhD student, they had repeated the poly-U experiment and were ready to embark on other synthetic RNAs, especially those with random mixtures of two bases. They soon established that poly-UC and poly-UA incorporated leucine into a peptide. But most of the progress was still being made by the better-equipped American laboratories: Nirenberg and Severo Ochoa's at New York University. Ochoa was the leader in synthesising RNA polymers and had therefore become the quickest to follow Nirenberg's lead. "The coding problem," Crick wrote, "has moved out of the realm of rather abstract speculation and into the rough and tumble of experimentation." Crick appointed himself the adjudicator of the many claims being made. He now had access to data from a set of German experiments on the tobacco mosaic virus, in which mutations caused by nitrous acid predictably changed cytosine to uracil and adenine to a chemical that behaved like guanine. This provided evidence at last that the code was degenerate: almost every triplet seemed to mean an amino acid, even if several triplets shared the same meaning. He could also check the results against spontaneous mutations in human haemoglobin, from which it soon became clear that the code must be the same in people, tobacco plants, and bacteria, thus confirming that it was universal.

Crick wrote a review, "The Recent Excitement in the Coding Problem," in which he handed himself the "thankless task" of assessing Ochoa's many somewhat hasty and untidy claims for the meaning of different triplets. Crick suspected that many were wrong, but a worse problem was that the cell-free system worked only with U-rich RNAs, because (in retrospect) the precipitating agent being used worked only with certain amino acids. His bold claim in the paper on triplets, that it would all be solved within a year, had by now proved vain. Nevertheless, he ended his review with another bold assertion: that the genetic code "is a non-overlapping triplet code, heavily degenerate in some semisystematic way, and universal or nearly so." Once again, he showed his knack for sifting the timeless truth from the contemporary confusion.

In February 1962 the Laboratory of Molecular Biology severed its links with the Cavendish and moved to a new building on the outskirts of Cambridge, next to the new Addenbrookes Hospital and a stone's throw from the Strangeways lab, where Crick had started his career in biology. At the same time, it absorbed Fred Sanger's group from the biochemistry department and Aaron Klug's from Birkbeck College. In all, about 60 scientists came together in the new minimal-modernist building. The informal tone of the hut—first names for everybody, no ties—continued. There were no committees or formal reports, and no grant applications. The scientists could study what they liked within the laboratory's budget. Perutz, ever the democrat, preferred to be called not director but chairman of a governing board, on which sat Crick, Kendrew, and Sanger. In May, the queen came to open the building formally. Crick and Brenner stayed away, saying that they disapproved of royalty. Watson, who was passing through England, happily took their place and chatted to the queen about horse breeding.

On 18 October 1962, three days after the Cuban missile crisis began, telephones rang in Cambridge, England; Cambridge, Massachusetts; and London. Crick, Watson, and Wilkins learnt simultaneously that they had won the Nobel Prize for physiology or medicine. The news was not a total surprise. The previous year, Jacques Monod had asked Crick for an account of the discovery of the double helix; he was preparing a brief—undoubtedly for the Swedish Royal Academy of Sciences. Crick had written back: "The data which really helped us to obtain the structure was mainly obtained by Rosalind Franklin, who died a few years ago." But Nobel Prizes are never given posthumously; nor is a single prize ever shared by more than three people. Had Franklin lived, the academy might have solved its problem by giving her and Wilkins the prize for chemistry. As it was, the chemistry prize went to Perutz and Kendrew for their work on protein, making a nearly clean sweep for Cambridge. The Russian Lev Landau won the prize for physics, but he had just suffered brain damage in an automobile crash and could not attend the ceremony in Stockholm. John Steinbeck won the prize for literature.

Odile was shopping in Trinity Street when she heard the news of the Nobel Prize from a passing friend. It was a Wednesday—early closing day for the shops—so she dashed to the grocer for food, the wine merchant for champagne, and the fishmonger for ice with which to fill the bathtub and cool the champagne. A party was bound to break out, and did, eventually spilling onto the roof of the Golden Helix. In the middle Watson telephoned. "I'm sorry if I was incoherent, but there was so much noise I could hardly hear what you said," wrote Crick the next week. Congratulations began to pour in, as did letters from cranks, autograph hunters, religious enthusiasts, and desperate invalids.

After popping over to Toronto to receive the Gairdner Award in November, Crick took all his family to Stockholm for the ceremony in December. His aversion to royalty did not apparently extend to the Swedish variety: he even bowed slightly on receiving the medal from the king. At the banquet after the ceremony Odile sat next to the 80-year-old King Gustav; and Francis—in white tie and tails—sat next to 24-year-old Princess Desiree. Steinbeck, Watson, and Kendrew made speeches. Crick sent his place card along the table to Watson with the note "Much better than I could have done, F." After dinner Crick was photographed dancing with 11-year-old Gabrielle. His lecture the next day was on the code.

At the turn of the year the BBC broadcast a television program, *The Prize Winners*, about the five molecular heroes. It was introduced by Lord Mountbatten. Angus Wilson's review in *The Queen* described Crick as:

> almost a caricature of the obsessively talking, idea-throwing-out don with a lot of disarming boyishness and bounce. . . . All the false hares and nutty suggestions, all the hours of exhausting listening and strained disagreement, are finally, miraculously made infinitely worthwhile when a man like Dr Crick eventually talks himself into one of the great revolutionary scientific theories of the century. How necessary . . . to have a combination of men like Dr Crick, who is too impatient for experiment and bubbling over with ideas, and of men like Dr Wilkins, whose life is given to experiment, who has infinite patience and a love of fiddling.

Bouncy or not, Crick did not much like his newfound fame. "Dr Crick never allows photographs of himself to appear anywhere if it can possibly be avoided," his secretary wrote to one fan in November. This was a restriction he managed to maintain for many years. In the years ahead he would refuse most honours that came his way, even honorary degrees. Watson says that Crick was never interested in seeing himself as a historical figure and disliked the obligations of fame. Apart from one set of drawings by Howard Morgan for the National Portrait Gallery in 1980, Crick never sat for his portrait. At some point he was tempted by the offer of a knighthood, and friends urged him to accept, since "Sir Francis Crick" sounded so much like "Sir Francis Drake." But Crick concluded that it was a useless bauble, intended to buy off scientists when what they really needed was better funding. He was not alone. The list of molecular biologists from Cambridge who won Nobel Prizes but refused a knighthood included: Crick, Perutz, Brenner, Sanger, Cesar Milstein, and Rodney Porter. Only Kendrew, Klug, John Walker, and John Sulston accepted; and Watson gained an honorary knighthood in 2001 after ensuring a large role for British laboratories in the Human Genome Project.

To deal with fame, Crick borrowed a joke from the American critic Edmund Wilson, and printed a card that read as follows:

> Dr F. H. C. Crick thanks you for your letter but regrets that he is unable to accept your kind invitation to:
>
> Send an autograph
> Provide a photograph
> Cure your disease
> Be interviewed
> Talk on the radio

Appear on TV
Speak after dinner
Give a testimonial
Help you in your project
Read your manuscript
Deliver a lecture
Attend a conference
Act as a chairman
Become an editor
Contribute an article
Write a book
Accept an honorary degree

(Some wag once returned it with "go to stud" added at the end.)

The days when Crick was thought a garrulous underachiever were now a distant memory. In 1961, along with Leo Szilard, Salvador Luria, Jacques Monod, and Warren Weaver, he was invited by Jonas Salk—who developed the first polio vaccine—to become a nonresident fellow of a new research institute that Salk was putting together in a futuristic building by Louis Kahn in La Jolla, in southern California, on land granted by the city of San Diego. This entailed annual trips in late winter at first to Paris and later to California, and eagerly anticipated bouts of argument with Monod as well as with the resident fellows, who included Jacob Bronowski and Leslie Orgel.

With his $17,000 share of the Nobel prize, Crick was also moderately well off for the first time. There was even time for a bit more leisure. In 1963 Crick at last learnt to drive—he was taught on a disused airfield by Odile in her Mini. In 1964 he bought his son Michael's MG sports car (originally a gift to Michael from Aunt Ethel when he graduated), and one day he and Odile happened on a pretty thatched cottage with a large garden in the Suffolk village of Kedington about 20 miles east of Cambridge. After persuading the farmer that it should be for sale, they bought that, too. There Crick took up gardening, throwing himself into the details of varieties of daffodils and roses with predictable thoroughness. In the autumn of 1964 he also bought a half share in a 47-foot Sparkman and Stephens yacht, *Kiwi 2*. It was jointly owned by the Italian scientist Giampero di Mayorca and was kept in Naples with a single elderly crewman, who spoke only Italian.

In 1965 Odile rented a villa in Capri from which they could sail while their artist friend Rodolfo de Sanctis took care of the two girls. But the expense of the boat was daunting, Di Mayorca was difficult, and Crick never became an accomplished sailor, so his share of *Kiwi 2* was sold a year later, to be replaced by *Eye of Heaven*, a Bertram powerboat that was delivered to Bari.

In Cambridge, Odile had begun exhibiting her art with a friend named John Gayer Anderson, who was known as the "squire of Waterbeach." He led a chaotic, polygamous life at his rambling house in the Fens, where wives, girlfriends and babies came and went, and where parties sometimes ended in unexpected pairings. Gayer Anderson, the son of a famous Egyptologist, was partial to parties; and when village life in Waterbeach palled, he would occasionally help to organise them at the Golden Helix. His sculptures were erotic, verging on pornographic, startlingly explicit even compared with Odile's many paintings of nudes. But the Cricks enjoyed his company and were not easily shocked, even when Gayer Anderson's 8-millimetre blue movies were projected backwards during a party. Entering into the artistic spirit, Crick even took up photography around this time, persuading Odile and sometimes the au pairs to pose for him. The 1960s had arrived.

A typical party given by the Cricks in the 1960s or 1970s, organised on the slightest pretext, would fill all four floors of the Golden Helix with friends, music on the gramophone (Mike Oldfield's *Tubular Bells* was a favourite), punch bowls of drink in the kitchen, and the scent of the odd joint in the air. Though they did not have an explicitly "open marriage," Francis was an incorrigible flirt, and Odile at least affected not to mind. He was apt to say to women at parties, "I know you are very happily married, but everybody needs a little excitement"; and he once told his startled secretary (apropos her recent marriage), "You can't expect a man to go through a long marriage without an occasional affair on the side." Yet he flirted in such a gallant and open way that few were offended and most were charmed. Sexual banter aside, secretaries who worked for him found him unusually warm, generous, and considerate as a boss.

Despite these distractions during the 1960s and the inevitable temptations to pontificate that come with a Nobel Prize, Crick's concentration remained focused on the lab. In the new, larger, laboratory, he took charge of making sure that everybody knew about everyone else's scientific work. Visitors from elsewhere in Britain and from abroad passed through regularly, always being made to give seminars; and "Crick week" was a week of seminars when the lab members told each other about their results. Sitting at the front, Crick was a terrifying presence, concentrating hard, interrupting frequently, and of course at the end giving a lucid summary of not only what the speakers had just said but what they should have said and what it all meant. At least one speaker was reduced to tears by the interrogation. Even those who asked questions were sometimes corrected: "The question you should have asked is . . . and the answer is . . ." Graeme Mitchison recalls that, as a result, the seminars were both a terrifying ordeal and an excellent spectator sport.

Crick was still preoccupied with the code. After the excitements of 1961, there was a pause. Beyond simple sequences like UUU, Nirenberg's cell-free system could not give unambiguous results when the precise sequence of bases in the artificial RNAs remained unknown. A new technique was required. Crick's main job was to prod and advise the chief experimentalists to test each other's ideas and explain anomalies. The breakthrough came in 1964, when Nirenberg and Philip Leder found that a ribosome would attach to a simple RNA triplet if the appropriate transfer RNA, with its amino acid, was also present. In addition, Gobind Khorana in Madison, Wisconsin, now worked out how to make messengers with alternating bases (such as UCUCUC). George Streisinger at the University of Oregon was analysing the effects of phase-shift mutations in phage proteins and getting results that seemed to support Nirenberg's. Charles Yanofsky at Stanford discovered in which direction the protein was built in bacteria—luckily it was in the same direction as the DNA was usually written—and found that single-letter mutations in a bacterial protein gave the same changes in amino acid as Nirenberg predicted. All the techniques were converging.

In January 1965 Crick arrived in America just too late to have lunch at a hotel in New York with Jim Watson, Salvador Dalí, and—as Watson said in a letter—"the most beautiful girl in the world" (it was Mia Farrow). But Crick had brought with him "tentative allocations for many of the 64 triplets." Travelling around the country on his way to the Salk Institute, and seeing how the various lines of evidence fitted together, he filled in several of the gaps in the list. At the Salk Institute he rented a house near the shore at La Jolla. One day he called Odile to say that they had been invited to go sailing, and asked her to make sandwiches. Odile found that she only had a frozen loaf of bread, so, intending to put it in the sun to thaw, she stepped towards the patio, and not noticing the glass door, crashed through it. Francis arrived home a few moments later to find her wearing only underwear, with her arms and legs covered in blood. He called an ambulance and she was rushed on a stretcher to Scripps Hospital, where a doctor sewed up her wounds with about 100 stitches. Afterwards, the Cricks always made sure that their glass doors were adorned with images of butterflies.

When Odile had recovered and they had travelled back to Cambridge, Crick turned again to the code, pulling together the threads he had learned from Nirenberg, Khorana, Streisinger, and Yanofksy. On 2 April, he first drew up the now classic table for the code, showing the first letter of each triplet in rows, the second letter in columns, and third letter as small rows within each row. In each cell was the name of an amino acid. The table was to become almost as iconic as the double helix. However, there were still 14 gaps. Crick sent this table to Watson and

Nirenberg, suggesting that he and they issue a "statement on the state of the genetic code." He went on:

> As you know, I have found myself involved in this, but as a collater of information rather than a producer. I am constantly having to provide copies of my private version of the code to interested people. . . . This year's Gordon Conference should provide an ideal opportunity, not to present a final version of the code, but the best version of most of it. . . . It is obvious that no one person has enough certain data to establish the code by himself, and also that by pooling all the information we can already arrive at most of the code.

Nirenberg replied: "Why don't you write it?"

In Cambridge, Crick's colleagues had made an exciting discovery. In 1963 Mark Bretscher had realised that the growing amino acid sequence remained stuck to its last transfer RNA unless the messenger was a mixture of A and U, in which case it detached. In other words, there must be a special triplet that signified "end of message" and that contained a mixture of A's and U's. Now Brenner rediscovered the same phenomenon genetically in a mutant phage called "amber" (the English translation of "Bernstein," the name of its discoverer). By analysing the amino acids in the protein, Brenner realised that the amber mutants always produced shortened proteins at a codon that was normally one letter away from the triplet UAG. He therefore hypothesised that amber mutants must contain the triplet UAG, which meant "end of message." A similar mutant, called "ochre," proved to be UAA. These two must be "stop" codons. (The "start" codon later proved to be the same as the one for methionine; all proteins start with methionine, but it is sometimes stripped off before use.)

Crick had begun to detect an interesting pattern in the code: the third letter in the triplet did not always matter. For example, CUX meant leucine, ACX meant threonine, and GGX meant glycine, where X could be any of the four bases. Crick set out to explain this by arguing that whereas the first two bases in each codon formed typical hydrogen bonds with their counterparts on the transfer RNAs, the third base formed a looser link, capable of "wobble." Robert Holley had just deciphered the sequence of a transfer RNA for the first time and was surprised to find that the triplet of letters that seemed to be the anticodon included, opposite the third position in the codon, the base inosine, a deaminated form of adenine that can pair with any of the other three bases. Crick pounced on this as further evidence that the third base might be designed to be flexible—to wobble. This allowed one transfer RNA to fit more than one codon, so different codons could stand for the same amino acid. (In bacteria there are about 40 different kinds of transfer RNA using 61 codons, thanks to wobble.) By the middle of 1965 Crick's "wobble" hypothesis was circulating among the cognoscenti in draft form. He concluded, in typical style, "It seems to me that the preliminary evidence seems rather favourable to the theory. I shall not be surprised if it proves correct." For some historians of genetics, the insight regarding wobble was final proof of Crick's genius: it was not a complicated point; but of the many people working in the field, Crick saw it first.

Just as earlier, with regard to the double helix, there was fractious competition between the laboratories working on codes. Marshall Nirenberg had fallen out with Heinrich Matthaei and soon fell out with Gobind Khorana, too. Crick unwisely stepped into the middle of the situation. In late April 1966 he accused Nirenberg of quietly "slipping out" a paper on the triplet-binding results in the *Proceedings of the National Academy of Sciences* to claim priority over Khorana. Crick wrote:

> I have, unfortunately, been involved in unpleasantness like this in the past [he was referring to collagen more than DNA], and I realise that one does not always see how one's actions will look to other people, and that one can make errors of judgment when priority is at stake. Nevertheless on the face of it I feel you owe both Gobind and me an explanation, if not an apology.

Nirenberg exploded at the accusation, pointing out that the results had been finished long before but he had not gotten around to writing them up. He said that Crick's accusation about seeking priority was "completely, absolutely wrong—wrong in every respect—and thoroughly unjustified." Crick climbed down, but not without a parting shot:

> Relax! Relax! I didn't really believe that you had acted in an elaborate underhand way, but I had to point out to you how it might appear to someone looking at it from outside. . . . You must realise that, rightly or wrongly, Gobind was upset when your PNAS paper appeared. . . .

> I do sympathise with your problem of writing up work already completed. Sydney and I often have the same trouble. In the long run I think it often pays to write things up promptly.

By early 1966 all but one codon had been cracked, the exception being UGA. On 5 May, Crick delivered the Royal Society's Croonian Lecture, setting out all the steps in the deciphering of the code and triumphantly unveiling the final chart, minus that one recalcitrant UGA assignment. Those who attended his Croonian lecture left with a sense of history in the making. Crick went on, a month later, to Cold Spring Harbor, where the mood of celebration and triumph continued. He opened the meeting with a fine speech, "The Genetic Code, Yesterday, Today, and Tomorrow," in which he hailed the great achievement of "linking the two great polymer languages" and ended, with a swipe at all those who had doubted him in the early days: "It will be difficult, after this, for doubters not to accept the fundamental assumptions of molecular biology." The end of the meeting coincided with Crick's fiftieth birthday, so Watson and the Harvard researcher Bob Thach drove to Entertainments Unlimited in Levittown and, from a book of pictures, chose "Fifi" to come out of a birthday cake on the balcony of Blackford Hall at the climax of the party. For once Crick was delighted to be "challenged by an alternative model," as one observer put it.

The possibility of a simple code at the heart of all biology, promised by the structure of the double helix, was now established fact. The little cipher chart was the secret of life—or would be if UGA would yield. Late in October it did. The work

SECOND POSITION

FIRST POSITION

THIRD POSITION

	U	C	A	G	
U	phenyl-alanine	serine	tyrosine	cysteine	U
					C
	leucine		stop	stop	A
			stop	tryptophan	G
C	leucine	proline	histidine	arginine	U
					C
			glutamine		A
					G
A	isoleucine	threonine	asparagine	serine	U
					C
			lysine	arginine	A
	*methionine				G
G	valine	alanine	aspartic acid	glycine	U
					C
			glutamic acid		A
					G

*and start

The Genetic Code, in the format drawn up by Francis Crick in 1966

in Brenner's lab that had cracked the first two stop codons had also uncovered some puzzling lethal mutations in phage that seemed not to revert, when further mutated, to anything. Then one day, Leslie Barnett tested one of these again—it was called "opal"—and found that when mutated with a mutagen that changed G to A, it reverted to ochre. Since ochre was UAA, this implied that opal was UGA (it could not be UAG, which was not a lethal mutation). Further experiments soon confirmed that UGA was a third stop codon—and it had fallen to Barnett, the quiet technician, to place the last piece in the jigsaw puzzle of life. Crick played no practical part in this discovery but found his name on the paper. When he asked why, Brenner told him, "For persistent nagging."

This time, unlike 13 years before, there was no eureka moment, and no victory for one team over another. There was only the completing of five years of hard, collaborative work following eight years of frustrated guesswork, but the result was in many ways as great an achievement as the double helix. Throughout the story, if Nirenberg had been the dominant experimenter, Crick had been the dominant theorist. As Judson put it, "By brain, wit, vigour of personality, strength of voice, intellectual charm and scorn, a lot of travel and ceaseless letter-writing, Crick coordinated the research of many other biologists, disciplined their thinking, arbitrated their conflicts, communicated and explained their results." But Crick was the first to admit that the genetic code was a triumph for experiment, not for theory. All that futile speculation, starting with Gamow and continuing with the comma-free code, had come to little. Crick was more than ever convinced that in science, theory must be the handmaiden of experiment.

Chapter Ten

Never in a Modest Mood

IN FEBRUARY 1966, Crick was invited to give a series of lectures at the University of Washington in Seattle, and these were published later that year as his first book. He wanted to call the book *Is Vitalism Dead?* but the University of Washington Press told him that nobody in America knew what vitalism was, so it ended up with a title that echoed John Steinbeck (Crick's fellow Nobelist in 1962): *Of Molecules and Men.* Crick's argument in the lectures was that the genetic code had extinguished all excuse for the belief that something other than mechanics and chemistry was needed to explain life. Shrugging off the achievements of Dalton, Wöhler, Darwin, and Mendel, each generation of vitalists kept insisting afresh that there was something mystical and irreducible at the heart of living things—Henri Bergson's élan vital being the most famous version. As late as 1958 an otherwise distinguished physicist, Walter Elsasser of the University of Maryland, had written a book based on his calculation that there was no room in a sperm or an egg for sufficient information to build a body, and that therefore something beyond chemistry was needed—something "biotonic."

Crick's aim was to scotch such nonsense. Describing the mechanism of the gene in chemical and physical terms, he answered his own question of 20 years before: "The borderline between the living and the nonliving does not cause us very serious trouble in explaining what we observe in terms of physics and chemistry." He realised that vitalists would simply fall back on claiming there was something special in the origin of life or in human consciousness—this was one reason why these subjects would be Crick's next targets. The last line of Crick's book is: "To those of you who may be vitalists I would make this prophecy: what everyone believed yesterday and you believe today, only cranks will believe tomorrow."

The book was well received. Arthur Kornberg wrote to say that he admired Crick's courage and candour in disposing of the latest crop of vitalists among physicists: "The supply from that source is astonishing." The veteran British geneticist Conrad Waddington took Crick to task for presuming that biologists could yet have confidence that consciousness would one day yield to the ordinary laws of chemistry and physics. Crick replied: "I think that consciousness or awareness will cease to be mysterious when we can describe the patterns of nervous impulse, in particular parts of our brain, and can show in a detailed way that certain patterns are associated with certain thoughts." This is the first mention of what he later called the "astonishing hypothesis."

On Crick's return from Seattle in March, Odile had an operation for a bunion, which resulted in serious complications. A series of pulmonary embolisms kept her first in the hospital and then in a nursing home for several weeks, till early April. And on 8 March, Crick's only sibling, Tony, died at age 47. Tony Crick had emigrated to New Zealand in 1948 because he disliked the nationalisation of the medical profession. Trained at Middlesex Hospital, he had seen action as a medical officer in the army in North Africa, Italy, and Greece before completing his training as a radiologist. In Auckland he became a well-known figure not only in the medical world but at the Royal New Zealand Yacht Squadron, where he kept his yacht, the *Princess Persephone.* One colleague wrote, "Clinical meetings were galvanised when his tall figure unfolded, and loose logicians scurried for cover." Larger than life, going straight to the heart of a question, with a gift for laughter—Tony had not a little of his brother in him. Francis had not seen him for many years.

Apart from the sadness of losing his only brother, Crick cannot have failed to notice that male Cricks did not live long—his paternal grandfather had died at 47, his father at 60, and his brother at 47. Perhaps thoughts of impending mortality contributed to the fact that in 1966, when Crick reached the pinnacle of his career, he also had his most bitter argument—with Jim Watson. In 1962, Watson had begun writing a book about the discovery of the double helix. He started at Woods Hole that summer, and was rather pleased by his very first line: "I have never seen Francis in a modest mood." He then put this book aside while he drafted a much praised textbook on molecular biology. He took up the book about the double helix again on sabbatical in Cambridge in 1965. Encouraged by the novelist Naomi Mitchison (with whom he stayed at Carradale, her house on the Kintyre peninsula in western Scotland) to tell the story as bluntly as possible, he soon finished a draft. Called "Honest Jim," it was a disarmingly frank account of the events of 1951–1953, deliberately seen through the eyes of a naive midwesterner and unsparing in its portrayal of the main characters, not least the author. There were echoes of Joseph Conrad's *Lord Jim* but even more of Kingsley Amis's *Lucky Jim*. Jim blunders his way to triumph in Watson's book, as in Amis's book, though in the former he does not get the girl. Nothing like Watson's "nonfiction novel" had ever been written before. Science writing normally portrayed discovery as a stately progress towards the truth by heroes rather than a competitive, error-strewn scramble by flawed human beings.

Watson sent a draft of the book to Crick in November 1965. That winter Crick was busy with coding; and, partly because of that and partly because he did not like what he saw of the book, he did not read it till he was prompted again by Watson in March. At that point he sent Watson a list of corrections and criticisms. For example: "You imply that the Fellows of Caius did not enjoy my company because of my laugh. I doubt if you have any evidence for this since in my early days at Caius I was quiet as a mouse." At this stage, though, Crick gave no sign of disapproving of the whole project. His tone was irritated, but calm:

> You have got the thing right in a sort of way, but . . . it is a distortion of the facts if one looks at it carefully. . . . Yours makes a good story, especially as it gives a rather

> vivid picture of what you were up to at the time, but what I miss in it is the intellectual conclusion that can be drawn about our work.

Crick admitted that he had himself given two lectures about the story of the double helix, though he had included "nothing like so much gossip."

Crick now set off for the Mediterranean. Wearing his Royal Navy cap but knowing almost nothing about engines, he, with Odile, took the new powerboat across the Adriatic from Bari and through the Corinth canal to the island of Spetsai, where he was to join a scientific summer school on molecular and cell biology that had been started by Marianne Grunberg-Manago. He then left his boat in Piraeus, intending to return in April 1967 and meet Monod at Samos—but he abandoned the plan when the Greek government was overthrown by a military coup.

Watson's second draft, now called "Base Pairs," reached Crick in September and produced a tirade about the title—"I do not see why I should [be] described as base"—but still no real objection to the book itself. Less than a week later, however, Crick wrote to Watson again, saying that he now wanted Watson not to publish at all. The book was "neither scholarly nor documented" but one-sided and naive, and it would set a dangerous precedent for scientific collaborations. In describing his sudden change of heart, Crick added: "I have always made plain to you my dislike of the whole idea of your book, and for this reason refused to read your earlier drafts." But what had really changed in the week between his two letters? The answer is that Crick had been to see Maurice Wilkins and had discussed the book with him. The two had incited each other to rage at Watson's draft. Lawyers for both now wrote to Nathan Pusey, president of Harvard, threatening legal action if Harvard University Press published it.

Watson replied to Crick's letter:

> I do not consider my book defamatory in the slightest to you. You have a strong personality which cannot be avoided if one is to write how you do science. In the early Cambridge days, there were people who thought you talked too much for what they considered your limited ability and insight. But as they were all wrong, I cannot see what harm it does to say that your amazingly productive career always did not have the support of everyone.

Watson had meanwhile cunningly asked Lawrence Bragg, who might be expected to dislike his portrayal even more than Crick, to write a foreword. This meant that Crick and Wilkins could not count on Bragg's support in urging Watson not to publish. Indeed, Bragg—who had faced a long struggle to emerge from his own father's shadow—rather liked the idea of helping Watson emerge from what he still regarded as Crick's shadow. Bragg wrote to Crick, admitting that Watson had "disarmed" him in persuading him that the book, though brash, was a "fascinating specimen of the impression Europe made on a young man from the States." Crick replied tartly that he had no objection to Watson's writing about his impressions of Europe, and went on to remind Bragg of the time in 1954, when Watson had not let Crick talk about DNA on the radio, and Bragg had told

Crick not to do so without Watson's consent. Presumably, therefore, Bragg should agree, now, that Watson should not publish without Crick's consent.

Crick was by now in a towering rage about the whole thing. All winter letters flew back and forth. In April 1967, when Watson's next revision (called "Honest Jim" again) arrived, Crick was far from being mollified. In fact, he exploded. His six-page letter, copied to the president of Harvard, was, Watson thought, far more defamatory than anything in the book:

> Should you persist in regarding your book as history I should add that it shows such a naïve and egotistical view of the subject as to be scarcely credible. Anything which concerns you and your reactions, apparently, is historically relevant, and anything else is thought not to matter. In particular, the history of scientific discovery is displayed as gossip. Anything with any intellectual content, including matters which were of central importance to us at the time, is skipped over or omitted. Your view of history is that found in the lower class of women's magazines. . . .
>
> One psychiatrist who saw your collection of pictures [Watson was a keen art collector] said it could only have been made by a man who hated women. In a similar way another psychiatrist, who read Honest Jim, said that what emerged most strongly was your love for your sister. This was much discussed by your friends while you were working in Cambridge, but so far they have refrained from writing about it. . . .

> My objection, in short, is to the widespread dissemination of a book which grossly invades my privacy, and I have yet to hear an argument which adequately excuses such a violation of friendship. If you publish your book now, in the teeth of my opposition, history will condemn you.

What was it that Crick disliked so much? It is implausible that Watson's portrayal of him really hurt. After all, despite the gibe about Crick's never being in a "modest mood," and despite plenty of mocking asides, Crick is the hero of the book—the unappreciated genius who earns the narrator's admiration and envy, and eventually seizes the prize. One or two remarks hinting at Crick's habit of flirting with pretty women might have irritated Odile, but in *Who's Who* Crick himself had listed his recreation as "conversation, especially with pretty women," so this was hardly news to her. The fact that Watson had chosen to write a book without Crick as coauthor may have rankled, but not to this extent. Nor was Crick motivated by guilt at the way he and Watson had used Franklin's data with insufficient acknowledgement. This was always more of a concern for Watson. (The phrase "Honest Jim" came from a researcher at King's, Willy Seeds, who had bumped into Watson in the Alps in 1955 and had asked, with heavy irony, "How's honest Jim?" before continuing down the mountain.) Watson's book had an element of confession of guilt: his first idea had been to publish it as a two-part article in the *New Yorker* under the title "Annals of a Crime." But Crick's attitude towards King's was unembarrassed: the discovery, not who made it, was what mattered; scientific results were not private property.

More likely, what Crick really minded about "Honest Jim" was, as he said, the cheapening of their achievement. For all his bounce, he was a man who believed in seriousness. In the 1940s, with Kreisel, he had turned his back on wit for its own sake. He saw himself as a dedicated seeker of great truths who had worked very hard, with long hours of reading, calculation, and intuition, to get to the point where he could make a great discovery; yet the world would now learn about the quest as if it had been just another soap opera. "It sounded as if anybody could have done it," Crick complained to Watson during a live broadcast on BBC Radio 3 some years later. Watson gave the impression that science was a game played in between parties and tennis matches. There is also a reactionary tone to Crick's comments. He wanted to see a scholarly, reverent account. Watson wanted to write an irreverent novel of the 1960s, warts and all. This contrast is well illustrated by Crick's later remark: "Watson's principal aim was to show that scientists were human, a fact only too well known to scientists themselves but apparently not, at that time, to the general public."

As the battle proceeded during the spring of 1967, Watson was gaining allies and Crick was increasingly isolated. Though plenty of people, including Perutz and Linus Pauling, disliked the book, only Wilkins manned the barricades with Crick; and, being Wilkins, he left Crick to do the battling. Bragg would not withdraw his foreword, though he did change it. Moreover, despite many small rewordings in the various drafts, Watson had cut very little—only a chapter about his adventures in the Alps in the summer of 1952 had gone. In literary terms, the book was going to be an undoubted triumph. It had strong characters, drawn from life with scandalous honesty, and an exciting plot building up to a dramatic denouement. John Maddox, the editor of *Nature*, pronounced the book "a valuable and sensitive account of the way in which interactions between people can influence the course of important events" and promised to ask Crick or Wilkins to review it: "That should be fun too." J. D. Bernal said that he could not put it down and described it, with heavy irony, as "a disgraceful exposure of the stupidity of great scientific discoveries."

Crick remained stubbornly defiant. In June he seemed to have won, when Nathan Pusey, the president of Harvard, ordered Harvard University Press not to go ahead with publication, because he did not want the university involved in an "international controversy among scientists." Joyce Lebowitz, the book's editor at the press, wrote to Watson regretting the "wretched decision" and suggesting an "icy ignoring of Francis Crick" at Cold Spring Harbor that summer, adding: "If the worst comes to the worst, give him one for me." But far from suppressing the book (now called *The Double Helix*), Pusey's decision was the making of it. Watson immediately went to a commercial publisher, the newly formed Atheneum, which sought legal advice on whether the book was libellous. (According to the lawyers, it was mostly not, though they did try, in vain, to get Watson to change the first sentence to "I can't remember ever having seen Francis Crick in a modest mood.") The book was published in February 1968. The reaction was, indeed, shock at the exposure of the human side of science, but also critical acclaim and commercial success. In the *New York Review of Books*, Peter Medawar called it a classic, noting perceptively that Watson's artless candour would excuse him, because "he betrays in himself faults far graver than those he professes to discern in others." *The Double Helix* has sold over 1 million copies.

The book's success gradually eroded Crick's grievances. For a while he and Brenner tossed across the room possible titles for a book that would represent Crick's revenge: "The Loose Screw," "Brighter Than a Thousand Jims," "Dr Virago." Crick even wrote an opening: "Jim was always clumsy with his hands. One had only to see him peel an orange." But his heart was not in it. He was never one to bear a grudge for long; nor was Watson. By the summer of 1969 Watson and his new wife, Elizabeth were staying with the Cricks at the Golden Helix; and three years after that, in August 1972, Watson and Crick were making a television programme for the BBC together, revisiting old haunts in Cambridge including the Eagle. Crick even admitted the virtues of *The Double Helix*:

> I now appreciate how skilful Jim was, not only in making the book read like a detective story (several people have told me they were unable to put it down) but also by managing to include a surprisingly large amount of the science.

Watson was not the only old friend to get the rough side of Crick's tongue during these years. Crick's former office mate Jerry Donohue, who now doubted that the double helix was right, drew him into a bad-tempered exchange about crystallography that ended with Crick remarking, "I have established to my own satisfaction that you do not yet have an adequate theoretical grasp of helical diffraction theory." In 1974 Alex Rich received a blistering accusation of plagiarism regarding the structure of transfer RNA that began, simply, "Does your name stink. Aaron [Klug] was convinced that once you had wheedled out the details of his structure you would attempt to publish it as your own. This is exactly what has happened." Six months and many long letters later Crick withdrew the charge of plagiarism and the correspondence petered out, Rich and Klug both still feeling very bitter.

Though now in his fifties, Crick embraced the spirit of the late 1960s. He wore sideburns, wide lapels, and loud shirts. In 1967 he joined the council of an informal organisation called Soma to campaign for the legalisation of drugs, and, along with 64 others, including Paul McCartney and Graham Greene, signed a full-page advertisement in the *Times* (of London) arguing for leniency for those convicted of possessing cannabis. Crick was certainly an occasional user of both pot and LSD. He was introduced to LSD about 1967 by Henry Barclay Todd, whom he met through Ruth Sheen, one of Odile's models. At a weekend in Kedington, Todd gave him some Swiss-made LSD, and Crick was fascinated by its effects—by how he became confused about what familiar objects were for, and by the way it seemed to alter the passage of time. He took it several more times but, contrary to rumours published later, never had any involvement in the drug's manufacture or distribution.

After LSD became illegal in 1966, Todd grew wealthy dealing LSD from two clandestine chemical laboratories until he was arrested and sentenced to 13 years in prison in 1977. Todd's main supplier, Dick Kemp, later claimed (to a friend who spoke to a journalist who published an article in a British newspaper after Crick's death) that Crick had once said he had been taking LSD when he discovered the double helix. This cannot be true, and not just because of the thirdhand source—the drug had been barely available in 1953; Todd is certain that he was the first person to give it to Crick; and neither Todd nor Odile recalls Crick's meeting Kemp. Nonetheless, the well-publicised arrest—in "Operation Julie" in 1977—of Todd and Kemp and their associates was an uncomfortable moment for Crick, who was by then living in America. He stonewalled all reporters inquiring about the subject.

Crick's libertarian views on drugs did not extend to other subjects. In 1966 his friend Noel Annan had left Cambridge to take up the post of provost at University College, London. Annan wanted to give Crick an honorary degree, but Crick would not accept one even from his alma mater. Instead Annan prevailed on Crick to give the Rickman-Godlee lecture on 21 October 1968. It was to be Crick's only public foray into policy issues, and it was not a success. Annan and the audience were quite shocked by some of his views.

No transcript of the lecture survives, but Crick's notes do. If they are accurate, he covered population, euthanasia, drug laws, and of course religion—a fairly typical agenda during the late 1960s. It is a shock to a modern generation to recall how much collective coercion people still thought acceptable in that decade. Paul Ehrlich's book *The Population Bomb* had just come out—to great acclaim, despite its apocalyptic and misanthropic tone. Crick's lecture notes sounded a similar tone: "Have people a right to have as many children as they please? The answer must be no—so how do we decide? Should thalidomide babies be allowed to live? . . . What deformities should be allowed and who should decide? (Since quantity more than adequate, why not increase the quality.)" These were not new, spontaneous thoughts. At a meeting of the CIBA Foundation on "Man and His Future" in 1963, to the dismay of Jacob Bronowski and Peter Medawar, Crick had said that he did not "see why people should have the right to have children" and that some form of licensing or taxing of children would soon be necessary to discourage breeding among people who were genetically less fit.

In the Rickman-Godlee lecture Crick was just as stark on the subject of death:

> When should people be permitted to die? . . . We cannot continue to regard all human life as sacred. . . . Should babies only be legally born when they are, say, 2 days old—i.e., have to pass an acceptance test by society. (We do this for motor cars—why not people?) Should we have "legal death" (like legal coming of age) at say 80 or 85? Doesn't mean you have to die then! Merely means that certain expensive medical treatment is no longer available to you.

As for religion, his notes read: "Christianity may be OK between consenting adults in private but should not be taught to young children."

Later, Crick regretted the Rickman-Godlee lecture. He told me towards the end of his life, "I think the UCL lecture was a bit rash. I realise you can't go about it this way: you've got to take account of people's sensitivities. And you have to get into ethical debates wholeheartedly if at all." So in the 1970s, when genetic engineering became the subject of an almost continuous ethical debate that smouldered on for the next 30 years, one voice you never heard was Crick's.

However, he did continue to dabble in the debate over genetic determinism, intelligence, and race. Like many biologists, Crick was dissatisfied with the perceived hegemony of nurture over nature, but his recommendation was drastic. In his notes for the lecture at UCL, he wrote:

> Acute need of information on the general assumption that education is all important. Nonsense. Need more studies on identical twins separated at birth. So why should not all twins be separated at birth? Adoption easy. (Not necessarily compulsory, but social pressure and financial inducement) or drugs to produce more twins.

In 1970, while reading Karl Pearson's life of the founder of eugenics, Francis Galton, Crick wrote to Bernard Davis of Harvard University repeating his call for parents of twins to be encouraged to "donate a twin." He added:

> My other suggestion is in an attempt to solve the problem of irresponsible people and especially those who are poorly endowed genetically having large numbers of unnecessary children. Because of their irresponsibility, it seems to me that for them, sterilization is the only answer and I would do this by bribery. It would probably pay society to offer such individuals something like

> £1,000 down and a pension of £5 a week over the age of 60. As you probably know, the bribe in India is a transistor radio and apparently there are plenty of takers.

A year later, following Arthur Jensen's famous article claiming that black people had an innately lower IQ than white people, William Shockley, inventor of the transistor, exasperated many fellow members of the U.S. National Academy of Sciences by his repeated demand for a big initiative to study the relative IQs of blacks and whites. When seven members of the Academy signed a statement criticising this, Crick objected and added his signature to a statement supporting Jensen and Shockley: "Resolution on Scientific Freedom Regarding Human Behaviour and Heredity." Writing to the biochemist John Edsall of Harvard, Crick said, "I think it likely that more than half the difference between the average IQ of American whites and Negroes is due to genetic reasons, and will not be eliminated by any foreseeable change in the environment. Moreover I think the social consequences of this are likely to be rather serious unless steps are taken to recognize the situation." He was even prepared to resign from the Academy over the matter. "I am sure you will realize that if the Academy were to take active steps to suppress reputable scientific research for political reasons it would not be possible for me to remain a Foreign Associate."

In reply Edsall protested that their objection was not to research but to an accelerated and politically motivated crash program of racial research. Indeed, Ernst Mayr, the evolutionary biologist, also of Harvard, who was another of the signatories of the original statement, wrote to Crick arguing that Shockley's focus on race was getting in the way of a more "positive" eugenics program, which he had long favoured but which was blocked by the demand for freedom of reproduction, "a freedom which fortunately will have to be abolished anyhow if we are not to drown in human bodies." Crick's reply contained the bizarre statement: "I myself do not feel very strongly either way about the Black-White distinction. If I have a prejudice it is against the poor, and in favour of the rich, but such an attitude is almost equally unacceptable to most people." He expanded on this point a few years later in a letter to Sir Peter Medawar: "I do not suggest that only the very rich or the very intellectual should have children (what a thought!) but roughly that upper and upper-middle class families be encouraged to have say 3 or 4 on average and manual labourers and obviously dim and disturbed people have 0 or 1." Medawar told him curtly that his project was an example of just the kind of utopian social engineering recently exposed and confuted in Sir Karl Popper's book *The Open Society and Its Enemies*. With that, Crick stopped trying to urge his eugenics programme on anybody.

"Nobel Prize Winner Crick Backs Jensen's Racist Theories" read a flyer produced to coincide with a lecture Crick gave in Seattle in 1973. But aside from this, what is remarkable about the long episode is how Crick avoided getting into public controversy while holding fairly strong views. Apocalyptic worries about world population were widespread in the 1960s and 1970s, and with them came the old temptation to worry about the deterioration of the species. Scientists who step into a political debate have often proved embarrassingly willing to allow utopian ends to justify collectivist, illiberal means.

Chapter Eleven

Outer Space

CRICK HAD NOW settled into a regular pattern. February would find him at the Salk Institute, soaking up Californian sunshine and scientific gossip, or sometimes in Marrakesh; July or August would see him in the Greek islands on his boat. *Eye of Heaven* was a cramped accommodation: Gabrielle and Jacqueline slept in the bunks, occasionally rolling on to Francis and Odile, who slept on the table between them. Francis loved fussing over the engine; but since he was as dextrous as a pound of sausages (Jacqueline's phrase), this was a lost cause, and he grew quite tetchy if teased. For revenge he would wake the crew at four o'clock in the morning, to get going before the wind rose. They anchored in deserted coves to snorkel, or tied up in harbours to eat at tavernas.

Crick usually combined these Greek expeditions with two weeks on Spetsai, at the summer school there. He had gone to the first of these meetings on Spetsai in 1966, mainly because he wanted to take the boat to Greece. But at that meeting he had been asked to "lend his name" for the organisation of the next meeting. The following year a coup by a junta of Greek colonels led to the cancellation of the summer school, which remained in abeyance in 1968, though this did not stop the Crick family from renting a villa in the hills above Hydra that year. In 1969 Crick found himself by default one of the main organisers. Some of his colleagues, especially in France, believed that Greece should be boycotted because the regime was dismissing academics and was torturing political prisoners. Crick felt otherwise, arguing that this would further isolate and harm innocent Greek academics, and that it was hypocritical to boycott Greece but still be prepared to travel to Madrid, Warsaw, or—a Crickian touch—the Vatican. Indeed, if police brutality was a criterion, he wrote to François Gros, "some of us might have doubts about coming to Paris."

In the end, the meeting of 1969 went ahead after Crick got the Greek government to agree to give visas to all invited participants, including 15 from behind the Iron Curtain, and not to send a government minister to address the meeting. Crick himself repaid the cost of the initial reception. ("You will notice that the main financial effect of this action was to transfer $500 from our pockets into that of the colonels," he wrote later to a colleague.) That October Crick and Jacques Monod drafted a rather vague letter to *Nature*, signed by them and several others, on how to handle such issues in the future: it provoked little response. The meeting for 1970 was postponed to 1971 and was eventually held not in Spetsai but at Erice, in Sicily. By 1972 the meeting was again held at Spetsai.

On his way back from Greece in August 1969 Crick stayed

with Jacques Monod at Monod's old family home in Cannes. Monod had delivered a series of lectures at Pomona College in California and wrote them up in English as a book, *Chance and Necessity.* He later rewrote the book in French, from which it was translated back into English. It was a philosophical polemic in favour of natural selection as the cause of life's diversity, and it would go on, when published in French, to inspire a generation of students while scandalising French intellectuals who were still wedded to various forms of evolutionary dirigisme. The book had a strong effect on Crick, who now saw clearly the differences between the physical and biological worlds. As he later wrote to Kreisel, "Because of the cunning shown by natural selection the whole of Nature is little more than a series of gadgets. This distinguishes [it] strongly from almost all the important problems in physics. Typically, the errors in one gadget are corrected in a further one."

In Cannes Crick spent a long morning discussing a draft of the book with Monod before joining him on his boat to sail to Corsica. The outward leg, with Monod's son and daughter-in-law aboard, was uneventful. On the return leg, Monod predicted that they would reach Saint-Tropez just in time to see the nightclubs open, but they soon ran into strong winds and heavy seas. Crick was alarmed to see Monod attaching himself by a line to the boat and gingerly inquired what to do if Monod fell overboard. They made it to port only as dawn was breaking and the nightclubs were closing. The next day, despite a broken engine and a strong northerly wind, they sailed along the coast to Cannes. Crick envied Monod's sailing skill, and perhaps Monod's polymath talent, too—"the scientist, the philosopher, the man of action and the musician," as Crick would put it in an obituary when Monod died in 1976. Though Monod was never a dyadic partner of Crick's like Watson or Brenner, he was someone Crick always admired. "Our friendship was not the friendship of those who were young together, nor were we intimate in the sense that we discussed our personal problems with each other. Rather, it was based, I think, on a steady admiration, seasoned with an affectionate recognition of each other's failings." And Monod once told Judson: "No man discovered or created molecular biology. But one man dominates intellectually the whole field, because he knows the most and understands the most. Francis Crick." Nevertheless, watching Crick nervously bring a boat into harbour, Monod once remarked that he had now seen Francis Crick in a modest mood.

Crick's finances had been much improved by a series of shrewd investments in property. He demolished Croft Lodge, the large house on Barton Road in Newnham where his mother had lived, and replaced it with a modern apartment building consisting of 20 flats. This was a risky but potentially lucrative venture. With characteristic thoroughness Crick took out a large loan, commissioned the development, supervised the architect, and set about selling the flats. Douglas January, Cambridge's largest real estate agent, was reportedly so impressed by Crick's business sense that he offered Crick a job. In 1967 Odile furnished three of the flats that remained unsold, in a modern Danish style; and Tom and Joan Steitz, visiting postdocs in the laboratory, were among the first to move in. Crick acquired another single flat at Quainton Close, off Newmarket Road, and rented that out.

Crick's main focus, after the code, was the organisation of genes on chromosomes in higher organisms (higher, that is, than bacteria). A problem was emerging from the mist: there was too much DNA about. As Crick summarised it in two lectures at MIT in 1972, human beings had about 1,000 times as much DNA as bacteria, and newts had many tens of times more than either frogs or humans. It could not possibly require 10 times as many genes to build a newt as a frog. Even in the small genomes of fruit flies, it was apparent that there was at least 30 times as much DNA as needed in each gene. "The major question is what is all this DNA for?" Was it "junk," an "evolutionary reserve," or something present to control the expression of genes?

At the summer school on Sicily in 1971, Crick came up with an elaborate theory that the coding sequences of genes themselves would prove to be on straight "fibrous" stretches of DNA, the pale "interbands" of chromosomes, whereas the control sequences would be found in the darker bands where the DNA coiled up into "globular" structures. Specifically, he thought that coiled "hairpins" of double-stranded DNA would twiddle out from the globules and at their tips would dissociate into single strands, the better to be recognised by proteins that controlled expression. The globular DNA scheme as a whole was to prove fundamentally wrong. It was one speculation too many. This humiliating failure reinforced the impression that the time had come for perspiration, not inspiration. Data, not theory, should be king. Besides, what had once been a cosy club of molecular biologists had become a far-flung industry, producing a voluminous literature that not even Crick could master.

Crick needed to change fields. While still doing molecular biology, he had already set out in two new directions, where thinking and analysing might still count. The first was the development of embryos from fertilised eggs. Sydney Brenner had recently chosen the nematode worm *C. elegans* as an experimental animal, partly with the aim of tracing development from egg to adult and reconstructing its nervous system on a computer. He and Crick set about recruiting some talented and independent scientists to work alongside them. Their recruitment techniques were somewhat unorthodox. Peter Lawrence, just back from a stint in the United States, gave a lecture in the Genetics Department, which Crick and Brenner gate-crashed, arriving late and causing much whispering. At the end they offered Lawrence a job. The mathematician Graeme Mitchison was summoned to an interview and asked three questions. The first was to identify an object Crick placed on the table. Mitchison said it was a model dog. It was in fact a model of the ethanol molecule. The second question was a joke designed to detect a sense of humour. The third was "Are you good with your hands?"—to which came the answer "I play the piano." Mitchison got the job.

Mitchison and Michael Wilcox were to study the ability of blue-green algae to differentiate roughly every tenth cell into a nitrogen-fixing heterocyst. Lawrence was studying the way hairs grew on the cuticle of insects, particularly the blood-sucking bug *Rhodnius*. The aim was to explore the idea that there is a gradient of some chemical that supplies cells within the embryo with "positional information" on where to make legs or arms or a head. Lawrence thought gradients might also be the key to polarity, the direction in which organs "know" they should grow. Crick immediately became interested in calculating how such gradients could arise from the simple diffusion of a chemical morphogen from a source cell. He was struck to find that the distance over which the gradients were supposed to stretch was about right, given what was known about diffusion rates through cytoplasm—his work at Strangeways on cytoplasmic viscosity was at last coming into its own. Crick drew in a mathematician, Mary Munro, to help with the calculations and demanded that Lawrence regularly print the results of his own experiments on the ripples in the hairs of *Rhodnius* and debate how to interpret them. Crick's obsession with gradients was to prove right, though it was not until the 1980s that the morphogens themselves—usually proteins, but sometimes messenger RNAs—were uncovered along with the genes that made them.

In the early 1970s, with the double helix approaching its twenty-first birthday, historians were beginning to take an interest. In 1972, the protagonists, except of course Franklin, were interviewed for a television program intended to be shown in both Britain and America. After much delay, during which the British schedulers said it was not scientific enough and the Americans said it was too scientific, the film was eventually shown as *The Race for the Double Helix* on BBC2 on 8 July 1974, with narration by the chairman of the BBC himself. He was a former biologist at Caius and a friend of Crick's from Strangeways days, Sir Michael Swann.

The same year Robert Olby's book *The Path to the Double Helix*, a scholarly account of the history and prehistory of DNA, came out with a foreword by Crick. In the foreword, Crick praised Olby for treating the science "more thoroughly and at a higher intellectual level" than Watson. Crick had collaborated closely with Olby's research for some years, even taking Olby to see his aunt Winifred in Northampton in February 1970. During the dispute over *The Double Helix*, Crick had been anticipating Olby's book, and at one point even suggested that Watson hand his account over to be included in Olby's book in some form. In 1968 Horace Freeland Judson, then mostly an arts writer for *Time* magazine, began a series of long interviews with the leaders of the revolution in molecular biology. Crick was away in Greece at the time, and Judson did not meet him until 1971. Judson found the scientists, like the Beatles and the Rolling Stones, wonderful to interview because unlike politicians they answered the questions and told you what they thought. He gradually compiled an extraordinary oral history of the subject; and after he moved to Cambridge in 1974, he used Crick as a reader of his early drafts. By the time his book *The Eighth Day of Creation* came out, it was a compendious and intimate portrait of the events of the 1950s and early 1960s, and it would become a classic in the history of science. In its cast of characters, Crick played the lead.

Meanwhile, in the lab, Sydney Brenner was suddenly obsessed with programming a huge computer in a windowless basement. The original plan had been to re-create the development of *C. elegans* as a software program, but increasingly the machine itself seemed to be the point. Colleagues had to devise stratagems to rescue each other from Brenner's interminable conversations about software. Crick had no interest in this, and though their badinage continued as before, their interests were now diverging.

Crick's other new direction at this time was the origin of life. His partner here was Leslie Orgel, who was now at the Salk Institute. Crick had begun theorising about the origin of the genetic code before it was completed, arguing that it looked like an arbitrary frozen accident that, once invented by a primeval organism, could not be changed because a change would produce many lethal mutations. That we were all descended from such an organism—that the code was universal or nearly so—now seemed to be certain. Crick was genuinely puzzled that no organisms with alternative codes appeared to have survived. After all, as he remarked, reptiles still lived in a world that contained mammals so why could not different creatures that used different codes hang on in the same way? At every step in the elaboration of the code, were all alternatives completely eliminated by competition?

Crick came tantalisingly close to a seminal notion that would become fashionable in the 1980s: the "RNA world." This is the argument that because RNA can both replicate information (as DNA does) and catalyse reactions (as protein does), so life-forms made of RNA had probably preceded modern forms made of DNA, RNA, and protein. He remarked in one paper that transfer RNA "looks like Nature's attempt to make RNA do the job of a protein" and that primitive machinery may have "consisted entirely of RNA."

In September 1971 Crick attended a conference at the Byurakan Astrophysical Observatory in Yerevan, the capital of Armenia, on communicating with extraterrestrials. There he joined a remarkable galaxy of scientific stars, gathered by Carl Sagan, including the cosmologists Tommy Gold and Frank

Drake; the physicists Freeman Dyson and Philip Morrison; the neuroscientist David Hubel; the artificial intelligence pioneer Marvin Minsky; the inventor of the laser, Charles Townes; the historian William McNeill; the anthropologist Richard Lee; and, from his own field, Leslie Orgel and Gunther Stent. It was a bizarre meeting, not least because, in the first thaw of détente, communicating with Russians seemed almost as exotic as communicating with aliens, though there was a brilliant simultaneous two-way translator, Boris Belitsky, to make it easier. One evening, as the toasts multiplied around the dinner table, Crick began to feel the worse for wear. Reaching out for a jug of what he thought was water, he poured a glass and drank deeply; too late he discovered that it was more vodka.

At the meeting in Byurakan, he saw his role as discussing the origin and nature of life, stressing to the assembled experts that life needed to replicate, to mutate, and to influence its surroundings. "Nature has this device of two languages, one of which is good for replication and one of which is good for expression, and has devised an extremely complicated apparatus to translate from one language to the other, the results of which are our genetic code."

Cogitating on the universality of the genetic code—with its puzzling implication of the uniqueness and improbability of life—and encouraged by the speculative mood in Byurakan, Crick and Orgel began to talk through an idea that would mature two years later as an article in *Icarus*, a journal of planetary studies. This article, "Directed Panspermia," argued, with commendably steady logic amid some giddily uncertain facts, that if life is improbable but the number of planets in the universe is vast, then life is likely to appear on some planet, but is also likely to reach an advanced stage there before appearing elsewhere. Members of an advanced life-form would eventually conclude that their own world was doomed, and that the best way to colonise other worlds across the great gaps of space was not by travelling themselves but by sending rockets containing simple bacterium-like life-forms. Since the universe is at least twice as old as Earth, there is a possibility, perhaps even a probability, that by the time Earth had cooled, some other civilisation had already reached this point and was already infecting our galaxy. Ergo, there was a chance that our common ancestor did not arise on Earth but arrived from elsewhere, deliberately sent by an intelligent life-form. The argument sounds nutty, and is, but then so are all theories about the origin of life. As for devising an empirical test, Crick and Orgel noticed that living creatures need molybdenum as a cofactor for several vital enzymes, and molybdenum is an extremely rare element in Earth's rocks compared with other elements such as chromium and nickel, which could do most of the same chemical jobs. Perhaps we all came from a molybdenum-rich planet elsewhere. Unfortunately for this argument, chemists soon pointed out that molybdenum is abundant in seawater.

For Orgel, the idea was a bit of a joke, but Crick tried to take it more seriously. His main motivation was to explain the universal code:

> It is a little surprising that organisms with somewhat different codes do not coexist. The universality of the code follows naturally from an "infective" theory of the

origin of life. Life on Earth would represent a clone derived from a single set of organisms.

But he knew it was a flimsy idea at best.

Embryology was too painstakingly empirical; panspermia was too ethereally speculative. Neither topic captured Crick, and by the mid-1970s he was back to the structure of DNA, suddenly fascinated by histone proteins that hold chromosomes together. Excited by the news that there were only five kinds of histones, the remaining 20 or so being postsynthetic modifications, he began to speculate about how DNA wraps itself around a combination of histones called the nucleosome, the realisation gradually dawning that the famous double helix is rarely straight in a chromosome, but nearly always curved. Aaron Klug and Roger Kornberg, who were doing crystallography on these objects, now found themselves bombarded with questions and suggestions. Kornberg was quite capable of interpreting his own results and was slightly surprised to find his crossword being finished by Crick in the usual way. It gradually became clear that chromosomes comprised a whole hierarchy of helices: the DNA double helix was itself wound around nucleosomes, which were packed end to end to make a larger solenoid, which in turn was wound around a still larger hollow cylinder, thus packing DNA into a space 10,000 times shorter than if it were stretched straight. Crick saw a general geometrical problem: how much and into what shapes you can twist a rope or a ribbon when coiling it up, a problem in topology that went under the name "writhing number." He had some fun with this. His great skill at visualising geometry, as in the days of helical diffraction theory, came well to the fore when he wrote a paper originally called "Writhing Numbers for Birdwatchers—an echo of the paper he had promised to write for Jim Watson in 1951. The new paper was eventually published in 1976 as "Linking Numbers and Nucleosomes." Crick even tried to twist rubber tubes into writhing supercoiled shapes. One day, using a half-closed window to hold a tube in place, he was tormented by bees coming in the window, little suspecting that one of the technicians was an avid beekeeper.

At times, though, some of Crick's zip seemed to have gone, and some colleagues suspected that he was slightly depressed during these years in the 1970s. At the end of 1971 he took two months off to recover from "overwork." In 1973, after a long trip to Washington State (where his son Michael with his new wife, Barbara, had recently settled), Hawaii, and Florida, Crick cancelled a visit to Pennsylvania and Tennessee and had to spend a few days in a hospital. He decided not to renew his Salk fellowship for a third six-year term and began telling people that he would cease accepting all invitations and would give up all unnecessary travel. He wrote to Watson in June 1974, "I myself am now heartily sick of the rat-race aspect of science but I still find the science itself absorbing." He did not even go to Spetsai that summer.

Part of the problem was his health. He was suffering intermittently from a painful complaint in his throat and chest. He would occasionally vomit blood. One day in 1975, while staying alone in a friend's house in London, he became very ill during the night. He called for an ambulance but almost had to crawl down a few steps to open the door. He was taken to Middlesex

Hospital, was diagnosed with a constricted oesophageal sphincter, and was operated on that morning—a long and complicated operation to stretch the valve at the entrance to the stomach. Odile, summoned from Cambridge to the hospital, waited anxiously for news for several hours. Crick was in intensive care for a few days and in a hospital ward for several more. Though he suffered from oesophageal reflux for some time and was briefly concerned about the possibility of cancer, he eventually made a full recovery and resumed his hectic schedule of travel as if nothing had happened. The spring of 1976 saw him at meetings in Switzerland, Turkey, Iran, and Germany; and that August he was back in Spetsai.

The obvious path to follow next—what most scientists would do at this stage of their careers—was to slide gently into grand panjandrumhood, becoming a professor; master of a college; or chairman of a government agency, research council, or royal commission of inquiry. Some bit of research could be kept open, the line being that "he still runs his own lab"; but the day job would no longer be discovery but management and politics. Watson had already taken exactly this step, giving up active research for administration and becoming the director and fundraising saviour of Cold Spring Harbor Laboratory. Crick was never tempted to do this—except once in November 1975, when he was approached by two fellows of Caius, the medical tutor Richard LePage and the poet J. H. Prynne, to consider succeeding Joseph Needham as its master. He agreed to let his name be put forward as a candidate, but after thinking about the proposal for a month, he and Odile decided that presiding over squabbling fellows, raising funds from rich alumni, and sipping sherry did not sufficiently appeal to him. He withdrew his name in January 1976. He needed to wake up each day and think about how the natural world, not the human world, worked.

Given the literary successes of Watson and Monod, it was evident that Crick should write a book. Indeed it was rather a surprise that he had not already written one, except for the pamphlet-length collection of lectures on vitalism, *Of Molecules and Men*. He had a fluent style, and his scientific papers were models of clarity. But he valued his own privacy too much to write about anything subjective, as Watson had done; and he valued empirical fact too much to fly off into philosophy, as Monod had done. Instead, he chose to write a popular book for a new publisher of illustrated books, Dorling Kindersley, which had already asked him to write the foreword for a children's science book. The subject Crick proposed was "scale"—the relative sizes of things from atoms to galaxies. By the end of August 1976 he had finished his first draft, "Travels with Francis Crick," and sent it off to the publisher.

The scare concerning his health, the approach of his sixtieth birthday, and punitive British taxation (even on foreign earnings) all came together in his mind to prompt the thought that he might emigrate, at least temporarily. In September 1975 the new president of the Salk Institute, Frederic de Hoffman, invited Crick to spend an eight-month sabbatical in California. Crick asked for unpaid leave from the MRC and began to explore what would happen to his pension if he took early retirement before his sixty-fifth birthday in 1981. To escape British taxation on fees from the Salk Institute, he needed to be employed abroad continuously for at least a year, so he planned three months as a visiting professor at Aarhus University in Denmark after eight months at the Salk and a month at Cold Spring Harbor. With her daughters now in their twenties, Odile was also free to consider living abroad. Though still based in Cambridge, Gabrielle was studying at Dartington College of Arts at Totnes in Devon; and Jacqueline was doing youth work in London. They joked to each other that since neither of them was ready to leave home, their parents had to do so instead.

Chapter Twelve

California

ON 10 SEPTEMBER 1976, the Cricks flew to California. There, they rented a house on Roseland Drive in La Jolla, bought a car, and took the state driving tests. Crick gave his first seminar at the Salk Institute, on chromatin, on 23 September and another soon after, on panspermia. The next year the Salk offered him a permanent position, handsomely endowed by the F. W. Kieckhefer foundation with a budget that allowed him to fly people from anywhere to the Salk for a few weeks of conversation. Thoughts of returning to Britain began to fade: the sunshine and prosperity of California, compared with the grimness of Britain in the 1970s, seemed irresistible. Also, as an employee of the Medical Research Council, Crick would soon reach the mandatory retirement age, and the notion of retirement horrified him. On 31 March 1977 he took early retirement from the MRC and became an employee of the Salk.

Crick's reddish hair was now going snow-white, and had mostly retreated to the back of his skull, but he sported generous sideburns and eyebrows as big as white mice. He was a little stooped now by years of trying to avoid bumping his head on door frames, but his six-foot-two frame was still slim and his blue eyes held a permanent twinkle of amusement. The Cricks bought a condominium in Solana Beach with the little money he could bring out of Britain, and rented the Golden Helix to students. For a few years they kept the cottage in Kedington for use during summer visits back to Britain, the first of which was from May to August 1978, but it was soon sold. Crick's decision to join the "brain drain" did not go down well with some of those left aboard the apparently sinking ship of Britain: there was muttered criticism in academic circles.

Meanwhile the book on relative scales had come to nothing. Felicity Bryan, Crick's literary agent, thought it very good, but spent the next six months apologising for Peter Kindersley's failure to come up with suggestions for how to rewrite it. Kindersley wanted something "far, far simpler," but his promised revisions never came, although in January 1977 a deadline of June was set. Crick was mildly hurt by this, not having had his prose rejected before. He was briefly tempted to write a book on Jacques Monod, who had died of cancer the previous year; and he then entered into negotiations with *Scientific American* to do a book on DNA for it; but he eventually decided against both projects. Instead, he began to realise that one of the ideas that sparked most with his lecture audiences was directed panspermia. So he sent the article he and Orgel had published in *Icarus* in 1973 to Felicity Bryan, and she sold the idea of a book to Alice Mayhew of Simon and Schuster. With remnants of "scale" thrown in as introductory chapters to make a coherent argument about the origin of life in the universe, the book—*Life Itself*—came out in 1981. It was a commercial success and was on the whole kindly reviewed, but the subject matter raised not a few eyebrows. The great Crick writing about alien life-forms seeding the universe from spacecraft? Had success gone to his head?

In 1978 he had agreed to write a memoir for a series sponsored by the Sloan Foundation and published by Basic Books. The first two books in the series—Freeman Dyson's *Disturbing the Universe* and Peter Medawar's *Advice to a Young Scientist*—had become best sellers. But after having signed up, Crick procrastinated; and it was not until 1986 that Sandra Panem at the Sloan Foundation managed to coax a draft out of him. Called *What Mad Pursuit*—the phrase from Keats that he had used in his first seminar on proteins at the Cavendish in 1950—the book recounted the main themes of his life with vigour and clarity. But Crick shied away from the story of the double helix, arguing that it had been too well covered already, and the text was predictably free of self-analysis. The reviews were polite but not ecstatic.

For some years Watson had been hinting at the possibility of a Hollywood movie based on *The Double Helix*. In 1981 the project got as far as a screenplay, and Watson and Crick both engaged agents to negotiate their fees as consultants. Crick was suspicious and cautious. Watson was much more enthusiastic, though he was appalled at the thought of being played by somebody as short as Richard Dreyfuss. The movie project petered out in 1984. Three years later, however, the BBC made a television drama, *Life Story*, based mainly on *The Double Helix* and starring Jeff Goldblum as Watson, Juliet Stevenson as Franklin, Alan Howard as Wilkins, and Tim Piggott-Smith as Crick.

Crick had intended emigration to mark a complete break from his old scientific topics and a new career studying the brain. But in 1977 a discovery in molecular biology drew his attention back to DNA. That summer Richard Roberts and Philip Sharp announced at Cold Spring Harbor that in animals and plants, unlike bacteria, many genes were split into stretches of sense, interspersed with stretches of nonsense, and that these nonsense stretches had to be excised from the messenger RNA before it was sent to the cytoplasm. The excised pieces soon acquired Walter Gilbert's name "introns," dividing up the "exons" of sense. Crick saw an opportunity to play his old role: to gather from the various papers a mass of new data, discard some, and stitch the rest into a tapestry that made sense. He wrote a long review paper for *Science* magazine, in which he speculated freely about the mechanism and the function of split genes and RNA splicing.

In the article, he mentioned an idea suggested to him by Leslie Orgel to the effect that some of the DNA in a genome might be "selfish DNA," which, by one mechanism or another, duplicates and spreads itself without doing much harm to its "host." Orgel had taken the idea from a paper by Richard Dawkins, who had first floated the notion in 1976 in his best seller, *The Selfish Gene*, as an explanation for the fact that most DNA was apparently not translated into proteins. Crick had now come to the view that the vast bulk of DNA was indeed "little better than junk" from the organism's point of view, but not from its own perspective: it consisted of sequences that replicated themselves like (mostly harmless) parasites. This is now generally accepted, and the notion of information parasites is now commonplace, thanks to computer viruses, unknown in 1980.

Although selfish DNA was to be Crick's last original contribution to the science of DNA, he was persuaded to become a consultant to one of the first biotechnology companies, Cetus, founded in San Francisco by Ronald Cape and Don Glaser. Cetus paid Crick $10,000 and some stock for four days of work a year. He spent much of his energy at Cetus damping down Cape's persistent enthusiasm for biochips that would combine genes with microprocessors, an idea Crick thought premature. He also suggested that they train bacteria to eat the "gunk" from coronary arteries and then isolate the enzymes as heart drugs.

Crick was now ready to carry out his long-held determination to switch his attention to the human brain. He had been thinking about the brain all his life. Consciousness was one of the two subjects he considered tackling, when he was preparing to leave the civil service in 1947, before plumping for life instead. To go back and, as an encore, knock off the other problem seemed only natural. He had known the physiologist Horace Barlow since the 1950s and had heard Barlow talk to the Hardy Club about "bug-detectors" in the frog's visual system and other clues to how the visual system worked. In 1964 Crick had been so fascinated by a seminar at the Salk in which David Hubel had described extraordinary experiments with Torsten Wiesel on the brains of monkeys that he had made Hubel talk for another hour. Hubel and Wiesel had found special brain cells that responded to specific features in the monkey's visual field—lines oriented at certain angles. Crick read all of Hubel's papers and continued to follow the work as the years went past. Then in

1972 he spent a week at a brain seminar at MIT, meeting many of the leading neuroscientists of the day. As soon as he moved to the Salk Institute in 1976 he began to immerse himself in the literature of neuroscience.

What he found was a field very much like genetics in the early 1950s: voluminous data, but with no core theories. In a manifesto published in *Scientific American* in 1979, "Thinking about the Brain," he wrote:

> It is not that neurobiologists do not have some general concept of what is going on. The trouble is that the concept is not precisely formulated. Touch it and it crumbles. The nature of perception, the neural correlates of long-term memory, the function of sleep, to give a few examples, all have this character.

The brain, like the gene before the double helix, was treated as a black box—you deduced things about it from its actions, not from its structure or mechanisms. Psychologists could gather good insights from such black-box work, but these could not be quantitative. "We must study both structure and function but study them within the black box rather than only from outside."

Psychologists reacted to Crick's condescension with a mixture of awe and irritation. A great man from a much more precise science had favoured them with attention, but he was implying that he could sort out their science for them—just as, in 1950, he had burst into crystallography and told the crystallographers that they were doing it all wrong. Compared with philosophers, the psychologists got off lightly: "Philosophers have had such a poor record over the last two thousand years that they would do better to show a certain modesty rather than the lofty superiority that they usually display." To his surprise, though, Crick found even most brain physiologists to be uninterested in the physical manifestations of thought. Those who called themselves cognitive scientists, for example, were eager to make theoretical models of mental processes and test how well they worked, but not to see if they were what real neurons were actually doing. Crick set out his wares as a critic of such "functionalists" and a champion of pure, reductionist materialism: the way to understand the mind was to understand its parts. He wanted to know not just what functions were being done by the brain, but "what sort of bits and pieces actually implement the functions under study."

He chose vision partly because of Hubel and partly because it gave such an accurate and vivid conscious picture of the external world with such deceptive ease, yet was so difficult to replicate by machine intelligence. He began by teaching himself the anatomy of the brain. Neuroanatomists were startled to realise that this was to be not the idle dabbling of an ageing dilettante, but the intense work of a student. Crick was prepared to attend seminars and lectures, read papers, and wade through experimental details. He wanted to see the subject from the bottom, not the top, to find the actual mechanisms of perception. He wrote: "We do not yet have any description of conscious perception that illuminates our very direct experience of it."

The first person he interrogated in depth was Valentino Braitenberg, a neuroanatomist based in Germany. They met when Crick came to Tübingen from Denmark in 1977 to give a talk. At a party afterwards Braitenberg, seeing Crick seated on a sofa surrounded by awestruck professors, sat down next to him and started talking about the anatomy of the fly's brain. Thrilled to find somebody interested in the concrete brain rather than the abstract mind, Crick asked to come to Braitenberg's laboratory the next day. "I have never met anybody who was able to absorb critically so many facts in such a short time," Braitenberg recalled. Later that year Braitenberg went to see the Cricks in Cambridge. He wrote to Crick in May 1978: "It filled me with joy to hear from you that you think, just as I do, that our main task now is to find the micro-circuitry underlying the Hubel-Wiesel etc. results." Braitenberg was invited to California for a month the following November, but he eventually failed to convince Crick of his particular model of the circuitry in question.

Crick's next target was David Marr, possibly the most renowned young brain scientist of the time. Marr, a mathematician by education, had risen to fame in Cambridge at the end of the 1960s, with a doctoral thesis in which he proposed a theory of mammalian brain function. After working at the Laboratory of Molecular Biology—where he was employed by Brenner in the latter's computer programming phase—Marr went into the study of vision and developed a revolutionary "computational" approach to perception. He argued that the brain must use mathematical algorithms to deduce features of the image presented by the eyes.

In April 1979 Marr brought Tomaso Poggio, a gifted Italian physicist who had worked in Tübingen on the visual system of the fly, to La Jolla to spend an intense month talking about vision with Crick. It was a thrilling time for all three, despite the shadow of Marr's recently diagnosed leukaemia, which would kill him the next year at the age of 35. A glimpse of their debates survives in an epilogue to Marr's posthumous book *Vision*, which includes a fictional Socratic dialogue between Marr and an anonymous sceptic. The sceptic is, more or less explicitly, Crick. In the dialogue Marr appears to convince the sceptic that understanding visual perception in the brain at the level of neurons alone will tell you nothing; a different level of computational explanation is also required. If Crick was convinced, he must have partly backtracked, because he remained a devotee of real neurons in the years to come.

As if to emphasise his reductionism, Crick first studied the structure of microscopic spines on the dendrites of neurons. Each pyramidal neuron, the commonest kind in the visual cortex, has about 10,000 of these tiny outgrowths, which make synaptic connections with the axons of other cells and through which most electrical traffic seemed to flow into the neuron. He became convinced by what he had read of them that dendritic spines literally twitched and that this twitching was central to their function, indeed that short-term memory might be stored in the pattern of advance and withdrawal of the spines. He predicted correctly that the spines had actomyosin in them, like muscle fibres. Nonetheless, it has eventually become clear that the movement is slower than he thought—a matter of tens of seconds, not fractions of a second. On a visit to Braitenberg and Poggio in Tübingen in 1980, Crick met their student who was modelling the electrical properties of dendritic spines. This was an American-born German, Christof Koch, who would become in the late 1980s and for the rest of Crick's life his final dyadic partner, after Brenner and Watson.

In the autumn of 1981 the Cricks embarked on a long speaking tour of Asia, returning to California by Christmas. On their return Crick lured his colleague at Cambridge, Graeme Mitchison, out to the Salk for two years to help him go through the existing literature on the brain. Graeme Mitchison—a nephew of Crick's old friend from Strangeways, Murdoch Mitchison; a grandson of Watson's muse, Naomi Mitchison; and a great-nephew of the geneticist J. B. S. Haldane—has a formidable mathematical brain (as well as an inexhaustible passion for climbing mountains, not something that had ever appealed to Crick). Perched in the first-floor corner office at the Salk that had once been Jacob Bronowski's, Crick sent Mitchison off to read up on the latest discoveries and to quiz the experimentalists in the labs downstairs, preparing himself for arguments with the master. Crick's reading habits were as diligent and thorough as ever, though he no longer came to work early. He would appear in mid-morning, then linger over lunch in the open air at the Salk canteen, usually with Mitchison and Leslie Orgel, as the hang gliders swept past over the cliffs. In their two years together Crick and Mitchison came up with two ideas: one on how a lattice of oriented axons could enable the brain to solve the problem of orientation detection à la Hubel and Wiesel; the other a theory of dreams.

The verdict of posterity on their theory of dreams, published in *Nature* in 1983, is mixed at best. The theory was based on the argument that simulated neural networks needed to purge themselves of unwanted or "parasitic" patterns of activity that arose spontaneously. If brains were the same, then dreaming might represent a sort of unlearning, a special mechanism operating during rapid-eye-movement (REM) sleep to seek out and eliminate these parasitic patterns and make them less likely in the future. "Put more loosely, we suggest that in REM sleep we unlearn our unconscious dreams." The theory suffered from a defect unusual in Crick's ideas: it was almost impossible to test.

Crick had now discovered two philosophers he liked: Paul and Patricia Churchland, both of whom had joined the University of California at San Diego, a short distance from the Salk. They chastised their fellow philosophers for continuing to argue abstractly about the mind as if neuroscience had never happened. The Churchlands' materialism was music to Crick's ears, and he persuaded the Salk Institute to make Patricia Churchland a fellow. Soon she and Crick were arguing happily together. "He showed us how to theorise," she said. "You actually had to create a detailed hypothesis, not just throw out a guess." Crick made sure they were joined by an ebullient psychologist, V. S. Ramachandran; and later by a brilliant expert on computational neuroscience, Terry Sejnowski. Never happy with yes-men, Crick was gathering around him the most talented people he could find.

In 1983 Crick, together with Ramachandran and Gordon Shaw, a physicist at Irvine, founded the Helmholtz Club, named after the nineteenth-century physicist who pioneered the study of vision. The purpose of the club was to debate the brain once a month. Crick became the central figure in its discussions. The meetings took place at Irvine, roughly halfway between La Jolla and Pasadena; they started with lunch, ran through the afternoon, and ended in a restaurant, with Crick's funds footing the bill. It was at one of these meetings in 1985 that Crick again met Christof Koch, who was soon to move to Caltech. He had already flown Koch out from MIT two years before for a week's interrogation (there is no other word for it) about the problem of visual attention, a subject on which both had just published papers. Now, after the club meeting, they began shouting happily across the restaurant table at each other about whether vision would be cracked in 10 years, as Koch insisted, or whether it would take longer. Crick saw something of his own younger self in Koch's intensity and self-confidence. Perhaps Koch was the partner he was looking for—Koch was a former physicist interested in real experiments and real brain parts. Without quite realising it, Koch was offered the job once held by Brenner and Watson: chief conversational partner, a role he would play for the next 18 years. The task Crick and Koch set themselves was consciousness itself.

Chapter Thirteen

Consciousness

CHRISTOF KOCH WAS BORN in 1956 in the American Midwest, and had lived in the Netherlands, Germany, Canada, and Morocco by the time he left school. After Tübingen and MIT he was appointed a professor at Caltech in 1986. Although, like the young Crick, Koch wears colourful waistcoats, in many ways they were very different. Koch indulges enthusiastically in running, rock climbing, science fiction, and the Roman Catholic faith, none of which appealed to Crick. But they had a genuine meeting of minds on neuroscience and found that they enjoyed arguing. Over the 18 years from their meeting in 1986, they would come to be very close friends. Koch often stayed at the Cricks' home—the Cricks had moved that year to a single-storey house with a garden and a pool on Colgate Circle, a quiet cul-de-sac high in the hills above La Jolla. He was allowed to raid the refrigerator at odd times of day. When Granada Television made an adaptation of the Sherlock Holmes stories, starring Jeremy Brett as the cerebral ultra-English detective, Koch was forcibly reminded of Crick and insisted that Crick watch it with him. Crick was unmoved.

Despite the difference in their ages, this was not a relationship between master and apprentice. Each might propose a theory, and each might write a draft—though imperceptibly, as the years passed, the initiative shifted to Koch and Crick stepped back into the role of adviser rather than instigator. (Crick gave a large photograph of himself to Koch shortly before he died with the legend: "Keeping an eye on you!") At the time they began to collaborate, scientists did not study consciousness directly; they left philosophers and cranks to speculate about it. No neuroscientist would dream of applying for, let alone getting, a grant to do an experiment on consciousness. "In our opinion," wrote Crick and Koch in 1992, "such timidity is ridiculous." Life had seemed just as elusive a concept as consciousness, before the structure of DNA emerged. Nor did it matter that consciousness was notoriously hard to define; so was "gene." The only sensible approach, they said, was "to press the experimental attack until we are confronted with dilemmas that call for new ways of thinking." Something somewhere in the brain is clearly happening differently when you consciously look at or imagine something. What?

It is hard to know how high Crick's hopes were when, in his seventieth year, he set out on this quest with Koch. He recognised that the analogy with DNA might be misleading. The reason there was a simple thing at the heart of life was that life necessarily started out simple: the double helix was left over from an earlier, simpler time. The brain, by contrast, was already a complex organ before it ever became conscious. Nonetheless,

Crick recognised many of the arguments that the "mysterians" used against his reductionism as ones that had already been used about heredity. Consciousness would be abstract, and dispersed, they said, and even when you found it you would not necessarily understand it any better. That was just what they had said about the gene.

He and Koch stuck to vision. Although to most people vision seems "unconsciously" automatic, this was a case of being misled by old-fashioned Cartesian dualism—the fallacy that there is a self, a soul, or a homunculus inside the head to which the eyes report. Simply to photograph the world with our eyes is not to see, for who then looks at the photographs? Somewhere in the brain there must be a representation of the visual world not in the form of an image but in the form of a symbolic "understanding" of an image. The self, in other words, is a shifting coalition of neurons themselves; and vision is an active, not a passive, process—the construction of an interpretation of what is received by the eyes.

Crick's favourite starting point was a simple visual illusion such as the so-called Necker cube, a line drawing that can be perceived as a three-dimensional cube seen from two quite distinct angles. It is possible, while staring at such a "multistable" puzzle figure, to have one's mind flip between the two interpretations, both of which are mental constructions. The diagram has not changed, but the conscious perception of it has. If one could find what had changed in the mind when such a simple transition occurred, one would be getting close to consciousness.

Crick and Koch called this the neural, or sometimes neuronal, correlate of consciousness (NCC)—the one pattern of brain activity, not necessarily in one location, that always coincided with conscious thought. Not that they would necessarily look for it themselves. They would comb the scientific literature, pick the brains of experimentalists, and nudge them in the direction of promising experiments. They began by making the point that you could clearly tell that some parts of the nervous system are not conscious. The retina, for example, is demonstrably unconscious of the fact that it has a blind spot where the optic nerve is. And as Nikos Logothetis proved with an experiment on an alert monkey, the first part of the brain's visual system, V1, cannot be conscious either. Each of the monkey's eyes was shown a different image, one moving up, the other down. Trained to respond in a different way to movement upwards as opposed to movement downwards, the monkey appeared to "perceive" an alternation between the two. But its primary visual cortex responded to what each eye saw, not what the monkey perceived.

This was the sort of hard experimental result Crick wanted, and he must have hoped that bit by bit neurophysiologists would in this way rule out different parts of the brain until they were left with a part that appeared to reflect conscious perception. But, though Logothetis eventually carried this technique farther until he could find neurons in a monkey's brain that responded to what it "perceived" rather than what it "saw," single cells were too uninformative. They proved only that conscious percepts are carried in brain cells, but not how or from where. Further clues came from neurologists' reports of the effects of brain damage. In 1986 Crick met the neurologist Oliver Sacks at a conference and asked him to "Tell me stories!" Each case history Sacks related set off bursts of hypotheses. "I never had a feeling of such incandescence," wrote Sacks afterwards. Some years later Crick was contacted by a neurosurgeon, Itzhak Fried, in Los Angeles, whose patients had chosen to undergo brain surgery for severe epileptic seizures. Koch then collaborated with Fried to record signals from single neurons in the brains of these patients during the surgery. They were able to find cells that responded to conscious percepts—for example, one in the amygdala that responded to three very different pictures of Bill Clinton but to no similar pictures of other presidents or other famous men. Clearly this cell was part of a network that held the thought "Bill Clinton."

By the early 1990s, Crick and Koch were excited about a suggestion by Christoph von der Malsburg that rhythmic, synchronised firing by neurons in the visual system of the cat might be a key to consciousness. This 40-hertz oscillation, perhaps in the pyramidal neurons, might be a way of "binding together" activity in different parts of the brain concerning the same percept. "Phase locked oscillations *are* the cellular expression of attention," they pronounced in 1991. Yet this was as close as they would come to a specific hypothesis of the NCC, and within a few years their enthusiasm for even this hypothesis had faded. In his book, *The Astonishing Hypothesis* (1994), Crick was more concerned to drive home the principle that consciousness must exist as a property of some neurons than to suggest exactly how and where. The book consisted mostly of an elegant discussion of the visual system of the brain, showing all his undiminished ability to gather facts and separate the grain from the chaff, but it also delivered a powerful polemic against the dualist notion that there is something immaterial or separate at the root of the self. It opens with a confident manifesto: "The Astonishing Hypothesis is that 'You,' your joys and your sorrows, your memories and your ambitions, your sense of personal identity and free will, are in fact no more than the behaviour of a vast assembly of nerve cells and their associated molecules." It ended with a rallying cry: "The case for a scientific attack on the problem of consciousness is extremely strong. The only doubts are how to go about it and when. What I am urging is that we should pursue it now."

The book's argument paralleled precisely that of Crick's first book and its attack on vitalism, the soul being a manifestation of the same urge. For Crick the reduction of the soul to an assemblage of neurons, far from removing mystery and awe, was a noble and uplifting quest—much better than clinging to the myths of the past. Those hoping that Crick might, in his seventy-eighth year, be showing signs of mellowing in his views of mysticism and religion, of reaching for the comforts of holism, or even of embracing Pascal's wager, were to be disappointed. Crick, like Darwin, would be a subject of rumours about a deathbed conversion, but the rumours were false in both cases. "The record of religious beliefs in explaining scientific phenomena [has] been so poor in the past that there is little reason to believe that the conventional religions will do much better in the future." This was the book of a man still fired by a youthful passion for truth.

The effect of his book was to make consciousness respectable. By trumpeting so loudly the need for a neuron-based attack on the problem, Crick had removed the need for neuroscientists to tiptoe around it. Although he accelerated this transformation, it might have happened anyway. Crick was not the only scientist from another discipline who had a go at consciousness in these years. In 1989 Gerald Edelman, a Nobelist in immunology; and Roger Penrose, a mathematical physicist at Oxford, both published books purporting to explain consciousness. Two years later the philosopher Daniel Dennett published a book actually called *Consciousness Explained.* Crick, who was not without some traits of a big beast asserting territorial possession, gave all three fairly short shrift. Edelman's theory of neuronal group selection he regarded as interesting but incomplete, and he described its author as "an enthusiast, noted more for his exuberance than for his clarity." Penrose's argument that a new form of physics, based on quantum gravity, would be needed to understand consciousness, he dismissed, but not before reading it thoroughly and corresponding with Penrose at length: "At bottom his argument is that quantum gravity is mysterious and consciousness is mysterious and wouldn't it be wonderful if one explained the other." Dennett, who argued that subjective consciousness was in fact an illusion, he described as "overpersuaded by his own eloquence."

The other big beast Crick fought with in the field of consciousness was Richard Gregory, a brilliant, pioneering psychologist, but one whose penchant for jokes and optical illusions perhaps marked him out as not serious enough for Crick. Gregory's playful use of analogies for what goes on in the brain was too much of the black-box school for Crick's liking. Crick made his disapproval known in no uncertain terms at one meeting of the Helmholtz Club, around 1990, heckling Gregory so continuously that for a while the seminar ground to a halt. Crick believed that analogies for brain mechanisms were pointless or worse; real neuronal phenomena must be described. Gregory responded that neuronal phenomena could not speak for themselves any more than electrical circuits could, and their function must be made clear on a conceptual level—a point very similar to the one made by Marr in 1979. On another occasion the two disagreed about why, when you look in a mirror, left and right are reversed, but up and down are not. The sight of two such eminent men arguing about something that most children expect their parents to explain caused much mirth.

This was an unusual case of Crick's politeness slipping. In neuroscience, as in molecular biology, he gained a reputation for courtesy, good humour, and even modesty. He might ask hard questions and press for hard answers, but his motive was to understand, not to win. He had great, immodest confidence in his ability to comprehend something—but he never believed prematurely that he had done so. Crick talked to everybody just the same: if somebody said something interesting, he had Crick's full attention; if he said something that exposed sloppy thinking, he would be told so in no uncertain terms. "I can tolerate somebody making foolish remarks for about 20 minutes," Crick once said: "Up to 20 minutes I can be very patient." Known to be interested in consciousness, he attracted endless letters and papers from cranks, new-age philosophers, theologians, and holists. He fended them off politely but firmly by saying that he could read only published papers.

In November 1991 Crick was invited by the queen to join Britain's Order of Merit, an honour limited to 24 members drawn from the arts and sciences. This time he accepted royal patronage because of its emphasis on "merit." Besides, though his dislike of a god was as strong as ever, his antipathy to royalty had mellowed. The order met every few years for lunch with the queen at Buckingham Palace. Crick and Odile went once, at last meeting the monarch he had avoided in 1962. Until the early 1990s, the Cricks would return to England every summer, eagerly asking Peter Lawrence which plays to see in London. The theatre was one thing Crick really missed in California, so sometimes when he was in London he would cram in two plays a day. But from 1994 on, he ceased travelling to England, though he did visit London briefly on the way to Germany in 1998. He and Odile bought a plot in the desert near Borrego Springs about 100 miles east of La Jolla in the early 1990s and, with him as architect, began to build a house. Desert gardening is a different skill, but Crick took to it with great zeal, gradually building a collection of plants that like dry conditions. He loved the desert light and had a special place from which to watch the sunset. If it was not too hot, he would occasionally walk the 3 miles to the Palm Canyon Oasis.

In 1995 Crick's granddaughter Kindra, the second of Michael's four children, spent a summer in La Jolla working at the Salk while she decided if biology would be her major at Princeton. She found her grandparents, now in their late seventies, "young at heart and fun." Francis Crick was always practical, always looking for a chance to laugh, always curious for new ideas. He encouraged Kindra to take drawing classes and to sketch with Odile. In the summer he would often eat lunch with friends by the pool, then swim at four in the afternoon. At dinnertime, since Kindra was not eating meat and Francis was not supposed to have butter or fat, Odile's legendary cuisine often gave way to expeditions to Indian or sushi restaurants, sometimes after the theatre or an art exhibition. Back at home, Crick would read till no later than ten at night, then close his book and announce that it was time for bed. Old friends and colleagues would sometimes come to stay. When Alison Auld, who had once been Crick's secretary and Odile's model (good looks were her chief qualification for both jobs), stayed with them, a casual conversation over dinner about her fascination with clairvoyance resulted in a pile of books by her place at the breakfast table, with relevant remedial passages neatly marked.

In 1994 the president of the Salk Institute had resigned unexpectedly, and Crick had taken on the position at short notice. He flung himself into the finances of the institute with enthusiasm, though he admitted that he did not enjoy doing this. Ill health gave him an excuse to step down after a year. He was diagnosed with heart disease and on 9 November 1995 underwent major surgery to bypass six arteries and replace part of the aorta. He made a good recovery, though he noticed that his moods were more variable. Any thought of "retirement" was quickly banished. He continued to read, talk, argue, and write as much as ever. With the distractions of frequent travel and administration now a thing of the past, his life had in a sense reverted to the pattern of the years before DNA. He inhabited a small universe of colleagues with whom he could debate scientific details all day and every day. He watched the completed sequencing of the human genome in the millennium year with detached pride at what had come from his insight, but took no part in the debates around it.

He guarded his privacy, if only to preserve his time for thinking. "I'm against communication," he joked, "because far many more people want to communicate with me than I want to communicate with." On the rare occasions when he did appear in public to give a lecture to students, his name drew huge crowds. While publicising *The Astonishing Hypothesis* in London in 1994, he filled the Methodist Central Hall in Westminster with over 2,000 people. He was also besieged by autograph hunters who got their way only if they donated ten dollars to the Salk Institute. Double-autographed copies of the original Watson-Crick paper had now become valuable items, and some people were sending reprints of the paper—cut from library volumes in a few cases—to both authors, hoping to get them autographed and then sell them. Watson and Crick agreed to put a stop to such profiteering by refusing to sign the copies.

In 2001, Crick reached a verbal agreement to sell the complete archive of his papers to a scientist, Al Seckel, who was working with a wealthy rare book dealer, Jeremy Norman, to collect scientists' private papers. But Crick was uneasy when he saw the written contract, which mentioned a third investor. His son Michael persuaded him not to sign; Jim Watson, hearing of the matter, quickly spoke to the Wellcome Trust and suggested that Wellcome offer to buy the papers. After a brief bidding war, Wellcome obtained a 50 per cent grant from Britain's Heritage Lottery Fund, and purchased all Crick's collected papers, which filled a dozen filing cabinets, for $2.4 million, with an agreement that copies would also be housed at the library of the University of California, San Diego. Seckel had offered more, but even this set an unprecedented price for the papers of a living scientist.

When cancer of the colon came in April 2001, Crick did not flinch. Christof Koch was with him when he took the call from the doctor confirming that the test results were positive. He put down the telephone, stared off into space for a minute or so, and resumed reading as if nothing had happened. Only later did he tell Koch what the call was about. There was to be no melodrama: his illness was just another fact about the universe. The doctors found themselves interrogated in detail about the treatments they prescribed, but to others he never laboured his medical problems.

Though no great breakthroughs came, Crick's thoughts on consciousness evolved during the late 1990s; and in 2002 he prepared what would be, if not the answer, his final framing of the question. Published jointly with Koch the following year, the paper was called "A Framework for Consciousness," echoing Crick's call many years before for neuroscience to get itself a framework before it could get a detailed theory, just as molecular biology needed a sequence hypothesis before it could crack a code. Crick and Koch set out 10 principles of which the central one was the idea of competing coalitions of neurons, the winning coalition somehow entering, or even embodying, "consciousness." There must, they said, be at least two kinds of coalitions, because the front of the brain is "looking at" the sensory output mainly from the back of the brain, rather like the mythical homunculus watching the screen. Attention, they argued, must be a mechanism for biasing the competition among coalitions of neurons. The NCC, with its "snapshots" of motion and its penumbra of semiconscious associations, may consist of quite a small set of neurons at any one time—perhaps only tens of thousands—and probably those that project from the back of the brain to the front. The framework was a hypothesis of sorts; but as Crick was the first to admit, it was far from detailed.

It was soon clear that Crick's cancer had spread, and chemotherapy was necessary. As 2003 approached, and with it the fiftieth anniversary of the discovery of the double helix, friends began to fear that Crick would not live to see it. But he did. Thanks partly to the Human Genome Project, and partly to the fact that DNA had become a household word (mainly, said Watson, because of O. J. Simpson and Monica Lewinsky), the fiftieth anniversary sparked far greater interest than had the twenty-fifth or fortieth anniversaries. Crick stayed well clear, sending videotaped messages to the main events. He granted only two published interviews at the time of the actual anniversary. In them he emphasised the futility of unreliable reminiscence. Far more important was what people wrote down at the time, he stressed. And the celebrations had gone too far: it was the molecule that mattered, not who discovered it.

Though frail and ill from chemotherapy, Crick continued to work as hard as possible throughout 2003 and even into the next year. He came to the Salk, or summoned people to his home. He had good days and bad days, but his eyes still let him read and his brain still let him think, so there was no reason not to work. The dining room table in the Cricks' home disappeared under neat piles of papers. Odile looked to his comforts, day and night, as she had done for 55 years, and to those of the stream of friends who called at the house. She or Kathleen Murray, Crick's assistant, took him to the clinic for treatments, where he insisted on walking, with painful slowness, into the building before returning to the car and demanding, with a wave of the walking stick: "Home, James, and don't spare the horses."

There was time for one more idea. In late 2003 Crick became obsessed by an obscure brain structure called the claustrum. He had known for many years that the claustrum, a thin sheet of simple neural tissue lying deep in the brain, is supremely well connected. It receives messages from, and sends messages to, all parts of the cortex and the thalamus. As he worked his way through the evidence that existed about the structure and activity of the claustrum, he gradually came to believe that it might be the source of a striking feature of consciousness: integrated unity. "You are not aware of isolated percepts, but of a single unifying experience," he wrote. "When holding a rose, you smell its fragrance and see its red petals while feeling its textured stem with your fingers." Yet smell, sight, and touch are processed in very distant parts of the brain. Something must integrate, synchronise, or "bind" them. The claustrum, with its multiple connections, its simple structure, and its uniform neurons, appeared to be in the ideal position to "bind disparate events into a single percept." But there was little evidence on what the claustrum was for. Being such a thin structure, close to so many others, the claustrum was almost impossible to knock out either by a stroke or deliberately, so it remained impossible to know what a brain without a claustrum would be like. Crick hoped that the newest molecular biology techniques for finding genes expressed only in certain bits of the brain would soon reveal some unique molecular signature of the claustrum, but he knew of nobody trying them. On Monday 19 July 2004, surrounded by piles of papers on the dining table at home, still mentally lucid but physically weak, he finished the first handwritten draft of a manuscript on the subject. Its last words were of characteristic urgency: "What could be more important? So why wait?"

A week later, on the afternoon of Monday, 26 July, Crick was taken to the hospital. On Tuesday he made corrections to a typed version of his draft about the claustrum. On Wednesday, 28 July, he worked a little more but then became semicoherent, imagining that Christof Koch was there and arguing with him about the claustrum. In the afternoon he sat with Odile, but she left the room when the doctors began a difficult procedure to ease his breathing. Half an hour later Francis Crick lost consciousness for the last time. A little after seven o'clock that evening, he died.

Epilogue

The Astonishing Hypothesiser

FRANCIS CRICK was cremated, and his ashes were scattered into the Pacific Ocean. On 3 August his family and his colleagues at the Salk gathered for a memorial at the institute, which consisted of a series of short eulogies from friends and family. On 27 September a more public celebration of his life was held at the Salk, in the open air on a hot, windy day as hang gliders scratched the sky behind the podium. Richard Murphy, president of the Salk, called Crick one of the greatest biologists of the century, if not of all time. Seymour Benzer, Leslie Orgel, Alex Rich, Aaron Klug, and Sydney Brenner each recalled working with Crick in the heyday of molecular biology. Jim Watson described him as sensible, pragmatic, never boring, unchanging, and "the greatest man I ever knew." From his later period, Tommy Poggio, Pat Churchland, V. S. Ramachandran, and Terry Sejnowski described his intellectual generosity and relentless logic in neuroscience. Christof Koch remembered his indomitable last days. And his son Michael posed the question

"What made Francis Crick tick?"—answering that he did not want to be famous, wealthy, or popular. He wanted to knock the final nail in the coffin of vitalism—a word, added Michael, that Microsoft Word does not recognise: "Score one for Francis!"

Because of the momentous nature of his discoveries, Francis Crick must eventually be bracketed with Galileo, Darwin, and Einstein as one of the great scientists of all time. Like them, he discovered a great truth that took the world by surprise—in his case the nature of life. Like them, he made many discoveries, not just one. Like them, he invented and dominated a whole new discipline. However, he did none of these things on his own. Does he deserve to join this pantheon for his genius, or was he just in the right place at the right time? He will always be best-known for being the first person, with James Watson, to see how DNA acts as a linear digital information-storage device, a completely unexpected result pregnant with enormous possibilities for the future of medicine, technology, and science. But in some ways this greatly understates his achievement, because he went on to discover that the code is a code for proteins, that it is spelled out in 64 three-letter "words" read sequentially along a DNA molecule, and that those words follow a universal cipher essentially the same in all creatures. Almost all the machinery of protein synthesis—the adaptor, the messenger, the triplet codon, wobble, stop codons, the sequence hypothesis, the central dogma, the cipher chart itself—bears the stamp of his (and Brenner's) insights and experiments. Today biology derives much of its new power from its ability to read DNA codes. And this is not even to mention Crick's contributions to protein crystallography, chromatin structure, embryonic development, and neuroscience.

That he did all this from a standing start, in early middle age, makes his youthful mediocrity all the more of a puzzle. Seen as merely clever by his teachers, he was widely thought a genius by his mature colleagues. What was it that made him so successful? It was not soaring mathematical flights (he turned to Kreisel or Griffith for these), or penetrating metaphysical complexities (for which he had no time), or even a persuasive fluency with words (though he was a good writer). His ability to visualise topology in three dimensions was remarkable and perhaps unique; but otherwise, by some lights there was something mundane and prosaic about his intelligence, something grounded in pragmatic, commonsense rationality, in guessing what "facts" to leave out and assembling the rest in a sensible pattern. Leslie Orgel describes him as "intensely intellectually organised." It was the intelligence of conversation and argument, not solitary inspiration. "Conversation was his grand stimulus," says Sydney Brenner. Nor did insight come to him with striking ease. He was certainly capable at times of quicksilver thought (Ramachandran says that Crick was "playful but passionate" in his approach to science), but he also had a prodigious appetite for doing his homework. Graeme Mitchison recalls "an inexhaustible tenacity that kept him worrying at a problem for long stretches of time." Aaron Klug was astonished by Crick's tolerance for reading even the dullest papers. Christof Koch often saw Crick read for two hours at a desk without a break. He told Seymour Benzer towards the end of his life that he used to be able to concentrate for eight hours, but now, in his eighties he could manage only six.

Francis Crick's genius was not of the kind that is close to

madness; he was not even eccentric. He trained his mind to be exquisitely good at solving nature's puzzles using logic, had the courage to take on the biggest problems, and threw himself exuberantly into the task, never letting prejudice stand in the way of reason. Throughout, he stayed true to himself: ebullient, loquacious, charming, sceptical, tenacious. He would have liked to find the seat of consciousness and to see the retreat of religion. He had to settle for explaining life.

Sources and Acknowledgements

TAKING FRANCIS CRICK'S own advice on the reliability of written rather than remembered evidence, I have leant heavily on the Wellcome Trust Library's collection of his papers. Some of his letters, lecture notes, and drafts of papers can now be found on the Internet at http://profiles.nlm.nih.gov/SC/. Crick's more than 150 published papers have not been assembled in any single volume, but drafts of many of them are in the Wellcome collection. As for published sources, the story of the Admiralty's magnetic and acoustic mines is well told in a memoir written by one of Odile Speed's bosses, Ashe Lincoln: *Secret Naval Investigator.* Robert Dougall's memoir, *In and Out of the Box,* tells of the expedition to Moscow. The history of molecular biology is found in Robert Olby's *The Path to the Double Helix* and Horace Freeland Judson's *The Eighth Day of Creation,* and of course in James Watson's *The Double Helix.* Other useful sources are Piergiorgio Odifreddi's *Kreiseliana*; Sydney Brenner's *My Life in Science;* Victor McElheny's *Watson and DNA;* Brenda Maddox's *The Dark Lady of DNA;* James Watson's *Genes, Girls, and Gamow;* and many other books. Crick's own books are *Of Molecules and Men, Life Itself, What Mad Pursuit,* and *The Aston-*

ishing Hypothesis. His ideas on consciousness are well reflected in Christof Koch's *The Quest for Consciousness*.

However, I have not relied only on written words. I sincerely thank the Crick family for all their help during the writing of this book. Odile gave me every possible encouragement as well as long interviews and replies to many written questions. Michael (and Barbara), Jacqueline, Camberley, and Kindra Crick all generously shared their memories of a remarkable father and grandfather.

Jim Watson showed me many early letters as well as being characteristically candid with his own recollections in several interviews. Georg Kreisel recalled the events of many years ago in several conversations and many letters. Christof Koch remembered his more recent close association in an interview and frequent E-mails. Sydney Brenner gave me a wonderful warts-and-all interview. Among Crick's other friends and colleagues who submitted to long interviews and persistent nagging by E-mail are Stuart Anstis, Michael Ashburner, Susan Blackmore, Valentino Braitenberg, Mark Bretscher, Pat Churchland, Raymond Gosling, Horace Judson, Sir Aaron Klug, Peter Lawrence, Graeme Mitchison, Leslie and Alice Orgel, V. S. Ramachandran, and Alex Rich.

I also thank Horace Barlow, James Barnett, Gerard Bricogne, Brian Dickens, Richard Gregory, Victor McElheny, Murdoch Mitchison, Kathleen Murray, Oliver Sacks, Tom Steitz, and Greg Winter for briefer conversations and exchanges. Among Crick's friends whose brains I picked to good effect were Alison Auld, Pauline Finbow, Sumet Jumsai, Dominic Michaelis, Robert Neale, and Nigel Unwin.

For various forms of practical help: in Northampton I thank

Sue Constable and John Peet; at Mill Hill Viv Wood; in Havant David Willetts and Betty Marshall; about the Admiralty Tim Lawrence, Stephen Prince, Angus Collingwood Cameron, and Pete Goodeve; in Cambridge Margaret Beeston and Sir Martin Rees; at the Wellcome Trust Helen Wakeley, Julia Sheppard, Richard Aspin, Tracey Tillotson, and Leslie Hall; at Cold Spring Harbor Jan and Fiona Witkowski, Bruce and Grace Stillman, David and Jody Stewart, Mila Pollock, and Maureen Berejka. Others who helped me in various ways were Steve Budiansky, Errol Friedberg, Georgina Ferry, Richard Henderson, Annabel Huxley, Jessica Kandel, Richard Le Page, Nikos Logothetis, Brenda Maddox, John McEwen, Tobi Megchild, Oliver Morton, Amanda Neidpath, Martin Packer, Tomaso Poggio, J. H. Prynne, David Roberts, Henry Todd, Sue Todd, and Christine Trimmer. For tracking down many articles and papers and photocopying them with great skill and speed I thank Paula McEwan. For printing and other practical help I thank Eunice Ridley and Jane Cowell. Robert Olby tolerated with good humor my intrusion into his subject. I thank John Kimball for permission to reproduce the image on page 70.

Felicity Bryan, my literary agent and Francis's, suggested that I write this book. I thank her and Peter Ginsberg for persuading me to do so, and James Atlas and Terry Karten for their encouragement and editorial wisdom. Odile Crick, Michael Crick, Anya Hurlbert, Christof Koch, Graeme Mitchison, Alex Rich, Jan Witkowski, and Jim Watson kindly commented on an early draft of the book.

I first met Francis Crick through my wife, Professor Anya Hurlbert, who worked with him in 1985. It is one of the many wonderful things she has done for me.